# Evolution Before Darwin

# Evolution Before Darwin
## Theories of the Transmutation of Species in Edinburgh, 1804–1834

Bill Jenkins

EDINBURGH
University Press

Edinburgh University Press is one of the leading university presses in the UK. We publish academic books and journals in our selected subject areas across the humanities and social sciences, combining cutting-edge scholarship with high editorial and production values to produce academic works of lasting importance. For more information visit our website: edinburghuniversitypress.com

© Bill Jenkins, 2019, 2021

Edinburgh University Press Ltd
The Tun – Holyrood Road
12 (2f) Jackson's Entry
Edinburgh EH8 8PJ

First published in hardback by Edinburgh University Press 2019

Typeset in 10.5/13 Sabon by
Servis Filmsetting Ltd, Stockport, Cheshire

A CIP record for this book is available from the British Library

ISBN 978 1 4744 4578 8 (hardback)
ISBN 978 1 4744 4579 5 (paperback)
ISBN 978 1 4744 4580 1 (webready PDF)
ISBN 978 1 4744 4581 8 (epub)

The right of Bill Jenkins to be identified as author of this work has been asserted in accordance with the Copyright, Designs and Patents Act 1988 and the Copyright and Related Rights Regulations 2003 (SI No. 2498).

## Contents

| | |
|---|---|
| *List of Figures* | vii |
| *Acknowledgements* | viii |
| 1. Introduction | 1 |
| 2. Edinburgh's University and Medical Schools in the Early Nineteenth Century | 7 |
|    The legacy of the Scottish Enlightenment | 8 |
|    The University of Edinburgh at the beginning of the nineteenth century | 10 |
|    The University of Edinburgh's medical school | 23 |
|    Edinburgh's extra-mural anatomy schools | 28 |
| 3. Natural History in Edinburgh, 1779–1832 | 37 |
|    Natural history in Edinburgh in the late eighteenth century | 37 |
|    Robert Jameson and the chair of natural history | 44 |
|    Comparative anatomy at the extra-mural medical schools | 54 |
|    Natural history, scientific and medical societies | 59 |
|    Natural history and science journals | 66 |
| 4. Geology and Evolution | 75 |
|    The Wernerian model of earth history | 76 |
|    Wernerians and Huttonians in Edinburgh | 79 |
|    The story of life as a tale of progressive development | 82 |
|    Wernerian geology and transformism | 86 |
|    Werner, Lamarck and Geoffroy in Edinburgh | 95 |
| 5. Edinburgh and Paris | 107 |
|    Contemporary transformism in France: Jean-Baptiste Lamarck and Étienne Geoffroy Saint-Hilaire | 108 |
|    Lamarck in Scotland | 120 |
|    The impact of Geoffroy's theories in Edinburgh | 135 |

6. The Legacy of the 'Edinburgh Lamarckians'  156
   The eclipse of transformism in Edinburgh  157
   *Vestiges of the Natural History of Creation*  161
   Transmutation without progress: Robert Knox and Hewett Cottrell Watson  177
   The legacy of Darwin's Edinburgh years  182

7. Conclusion  196

*Bibliography*  202
*Index*  219

# *Figures*

| | | |
|---|---|---|
| 2.1 | John Barclay attempts to enter the University in 1817 astride the skeleton of an elephant | 24 |
| 2.2 | Line engraving of Robert Knox | 31 |
| 3.1 | Robert Jameson | 45 |
| 3.2 | Robert Jameson through the eyes of one of his students | 47 |
| 3.3 | An undated print showing Edinburgh College Museum as it would have appeared in Robert Jameson's time | 51 |
| 3.4 | Lithograph of Robert Grant by T. Bridgford | 57 |
| 5.1 | 'Presumed order of the formation of animals in two separate series', from Lamarck's *Histoire naturelles des animaux sans vertèbres* | 112 |
| 5.2 | 'The origins of the main subdivisions of the animal kingdom', from Lamarck's *Philosophie zoologique* | 113 |
| 6.1 | Diagram showing the process of differentiation during development, from *Vestiges of the Natural History of Creation* | 171 |

# *Acknowledgements*

First and foremost I would like to thank John Henry, who supervised the PhD research on which this book is based, for all his help, support and wise advice during my doctoral studies at the University of Edinburgh. I would also like to thank my second supervisor, Thomas Ahnert, for much help and guidance. The final form of this book owes a great deal to Pietro Corsi, who was my external examiner and provided many invaluable insights. It is still a mystery to me how he apparently effortlessly identified the source of an anonymous article from the *Edinburgh New Philosophical Journal*, about which I had spent far too much time fruitlessly speculating. Niki Vermuelen of Science, Technology and Innovation Studies at the University of Edinburgh gave valuable support and kept me in gainful employment during the writing of this book. The encouragement and kind words of Jon Hodge and Ben Marsden regarding my work sustained the conviction that what I was doing was worthwhile at crucial moments. Aileen Fyfe also provided some good advice while I was working on the final revision of the manuscript. Special thanks are also due to Morag Ramsay who checked all my translations from works in French and saved me from one or two embarrassing errors. My wife, Catherine Laing, read almost every word of my thesis as it was written and provided many extremely valuable insights for which I am eternally grateful.

I would also like to thank all the staff, sadly too numerous to name individually, at the following libraries, who helped me locate the sources, published and unpublished, that formed the basis for my research: the Centre for Research Collections at Edinburgh University Library; Special Collections at Glasgow University Library; the Manuscript Collection at the National Library of Scotland; Archives and Manuscripts at the Wellcome Library in London; and the University College London Special Collections reading room at the National Archives. I would also like to thank Sue Beardmore at Elgin Museum, for tracking down George Gordon's notes from Robert Jameson's lectures for me, and Leonie Paterson of the Royal Botanic Garden, Edinburgh, for guiding me through the archives of the Royal Caledonian Horticultural Society. Special thanks are due to Elizabeth Singh and Enid Gardner of the Royal Medical Society for giving me access to the archives of the Society and assisting me in my extremely fruitful research among their wonderful collection of dissertation books.

My gratitude is also due to John Watson, Jenny Daly, Adela Rauchova, David Lonergan, Eddie Clark, Rebecca Mackenzie and Sarah Foyle of Edinburgh University Press for their help and advice during the preparation of the manuscript and for seeing the book through to press. Belinda Cunnison, EUP's copy-editor, also deserves my thanks for her work on the manuscript. Finally, I would also like to thank the two anonymous reviewers for Edinburgh University Press, without whose extremely helpful comments and suggestions for improvements this book would be very much the worse.

*In memory of my parents, Bill and May*

# 1

## *Introduction*

In October 1825 a young man from Shrewsbury arrived in Scotland's capital to study medicine at the city's famous university. It was no surprise to find him here. He was merely following in the footsteps of his father, uncle and grandfather, who had all studied medicine at Edinburgh before him. However, unlike his father and grandfather, he was never to graduate from the University of Edinburgh, but would leave the city two years later without completing his studies. That young student was of course Charles Darwin. When he arrived in Edinburgh Darwin found himself at a university that was very different from the University of Cambridge, from which he would eventually graduate with a Bachelor of Arts degree in 1831. The students at Edinburgh had enormous freedom to direct their own studies. They paid individual professors directly for their classes, which they could take in any order they chose, even if certain classes were obligatory for those who wanted to graduate. This 'market' in education meant that professors had to take their teaching responsibilities very seriously, as many of them, particularly in the Medical faculty, received no remuneration for their teaching other than class fees. It therefore paid them to make their classes both entertaining and enlightening. Some professors made significant sums of money by attracting large numbers of students who had no intention of graduating but attended the lectures purely out of interest or for the practical knowledge they could gain. In contrast to the situation at the the two English universities that existed at the time, the vast majority of professors were laymen rather than clerics. They were appointed for life and had almost complete freedom over what they taught and how they taught it. Many of them engaged in lively debates and disputations with their students and encouraged them to engage actively with their subjects rather than being mere passive recipients of knowledge passed down in lectures. Robert Jameson (1774–1854), the professor of natural history, whose lectures Darwin was to attend, was a model of this approach to teaching.

The student body also had a very different profile from that found at contemporary Oxford and Cambridge Universities. R. D. Anderson has noted that '[f]ew countries have had universities as confined to the elite as England in 1800'.[1] The situation was very different at the Scottish universities. Very few Edinburgh students were children of the aristocracy. The majority were sons of the middle classes, although they were surprisingly socially diverse, including a significant number of students from more humble backgrounds. Many of the lecturers made special allowances for such students, even on occasion waiving fees for promising students of limited means. Medicine at Edinburgh attracted many English students from non-conformist families, who were denied an education at Oxford or Cambridge by the requirement that students subscribe to the Thirty-Nine Articles of the Church of England. Students at Edinburgh were given much more freedom and responsibility over their studies and their own welfare during their time at the university than their contemporaries at the English universities, where the collegiate system created a strongly paternalistic atmosphere. Unlike Darwin, who seems not to have responded well to the challenges of student life at Edinburgh and the rigors of medical training and was unhappy there, leading him ultimately to abandon his studies, many students thrived in this stimulating and challenging environment and looked back on their student years with pleasure from the vantage point of their subsequent careers.

Some of the most exciting thinking and teaching in Edinburgh at this time was going on not in the University but in the city's extra-mural anatomy schools, which were privately run institutions and did not belong to the University. Many of the teachers at these schools, such as Robert Knox (1791–1862) and Robert Grant (1793–1874), were immersed in the latest continental theories and were keen to spread these ideas among the numerous medical students who attended their classes. The bold and provocative lectures of Knox seem to have made him immensely popular with students thirsty for challenging new ideas, and Grant was the centre of a circle of medical students who discussed the latest theories during long field trips to the Firth of Forth. In the earlier part of the period, at least, both Knox and Grant were also important figures in the wider natural history circles around the University's professor of natural history, Robert Jameson.

Edinburgh had been remarkable for its respect for freedom of thought and expression since at least the last decades of the previous century. At that time the University was dominated by a circle of key figures of the Scottish Enlightenment who, while, not necessarily sharing their views, had rallied round in defence of the rights of their friends David Hume

(1711–76) and Henry Home, Lord Kames (1686–1782), to express ideas that many considered dangerously radical. William Robertson (1721–93), principal at Edinburgh during those years, had a particularly profound influence on the culture of the University. Responding to this freedom, the students at Edinburgh, and medical students in particular, became well known for engaging with the most challenging new scientific and philosophical doctrines. The many student societies provided fora for fearless debates on subjects that would have been taboo at the English universities, including materialist theories of the human mind, the spontaneous generation of life and the transmutation of species.

It should come as no surprise that students at Edinburgh were discussing the transmutation of species in the early nineteenth century. It would perhaps be more surprising if they were not. The subject was a hot topic across Europe and lively debates on the subject were taking place from Paris to Göttingen. Through its close contacts with continental thinkers Edinburgh was subject to the same currents of thought as the other major European centres at this time. Many of the key figures in Edinburgh natural history had travelled to sit at the feet of continental masters such as Abraham Gottlob Werner, Georges Cuvier and Étienne Geoffroy Saint-Hilaire and study at the great European centres of learning. The open and tolerant atmosphere of Edinburgh and the numerous opportunities for students to exchange and discuss the latest ideas, both in class with their professors and among themselves in the many thriving student societies, provided an ideal environment for the propagation of exciting new perspectives on the natural world. Seen in a global perspective, it is perhaps Oxford and Cambridge that should be seen as parochial backwaters at this time.

Some scholars, notably Adrian Desmond in his valuable study of medical circles in Edinburgh and London, and Steven Shapin and Roger Cooter, in their important work on phrenology in the early nineteenth century, have portrayed the medical and natural history circles of Edinburgh as riven with conflict between Whigs, Tories and Radicals, each bringing their distinct ideological perspectives to bear on issues such as the transmutation of species.[2] 'Radical' ideas in natural history, such as evolutionary speculations, are seen as the province of political radicals. However, examining the careers of the key transformist figures in Edinburgh in the early decades of the nineteenth century throws up some surprising connections, which are very hard to explain under this model. As we will see below, an establishment figure such as Robert Jameson could be an important patron for both of the key Edinburgh 'Radicals', Robert Grant and Robert Knox. Grant could also be a personal friend of the Evangelical minister John Fleming, who even named a new species of one of the sponges Grant studied in his honour. The minutes of the Royal Society of Edinburgh show that at its meetings the arch-Tory Sir Walter Scott, president of the Society from 1820 to 1832, could rub shoulders with Knox, who frequently read papers to the Society during Scott's presidency. There was conflict between individuals, certainly, but in this study I contend that these can more often be explained in terms of personal rivalries and animosities rather than as manifestations of the deeper conflicts between social groups or classes that did undoubtedly divide Edinburgh society in those years. All the evidence suggests that political and religious differences counted for little in Edinburgh natural history circles in the 1820s, and were set to one side when discussing the great question of the nature and origin of species. In a more recent work, Shapin has proposed that 'we cannot understand how various scientific and technological knowledges are made, and made authoritative, without appreciating the roles of familiarity, trust, and the recognition of personal virtues'.[3] It was primarily their mutual respect as fellow natural historians that held Edinburgh's natural history circles together. Like the members of the early Royal Society of London described by Shapin in his *Social History of Truth*, the Edinburgh natural historians inhabited a 'discursive culture which was elaborated by the general desire to prevent dissent from provoking disaster'.[4] My focus, then, will be on the ideas that brought them together rather than the political or religious allegiances that may have divided them.

A number of important scholars, including James Secord and Jonathan Hodge, have long been of the opinion that evolutionary theorising was prevalent in Edinburgh natural history circles in the 1820s, when Darwin was a student there.[5] Open any scholarly biography of Darwin and you will learn about his membership of the Plinian Natural History Society, where controversial ideas such as spontaneous generation, materialistic theories of the mind and the inheritance of acquired characteristics were openly discussed. You will also learn of his friendship with Robert Grant, probably the most important promoter of evolutionary ideas in Great Britain in the early decades of the nineteenth century. Based on these facts and hints from other sources, James Secord has suggested that there existed a group of followers of the French evolutionary thinker Jean-Baptiste Lamarck (1744–1829), the so-called 'Edinburgh Lamarckians', around the better known figure of Grant and that this group may even have included Robert Jameson.[6] Secord considered it likely that an anonymous article entitled 'Observations on the nature and importance of geology', published in the *Edinburgh New Philosophical Journal* in 1826, was written not by Grant, but by Jameson. In this intriguing paper the anonymous author expressed admiration for Lamarck and his transformist theories, which he went on to discuss at some length.[7] Secord has suggested that, in the light of his reassessment of this article, 'our current picture of the acceptance of evolution needs to be overhauled'.[8]

My aim in this book is not to present the evolutionary thinkers active in Edinburgh in the first half of the nineteenth century as in any sense 'precursors' of Darwin, but rather to place them in their own unique historical and geographical context. Likewise I will generally avoid describing their theories as theories of 'evolution', as this term came into wide use with this meaning only with Darwin. To make this distinction clear, I prefer to use the terms 'transformism' or 'transmutation of species' rather than 'evolution'. I will do this primarily to avoid confusing the ideas I am discussing with the later Darwinian theory of evolution, which has significant differences from the earlier ideas I wish to address. Transformism, taken from the French *transformisme*, was not in use in the 1820s or earlier, but was widely used in the later decades of the century to designate non-Darwinian theories of evolution. The word 'evolution' itself was in fact fairly widely used in biological discourse in the eighteenth and early nineteenth centuries, but in the very different context of the 'unfolding' of the developing foetus.

While Darwin was present in Edinburgh for part of the period under study, and it would be wrong to ignore the evidence of the recollections he later wrote of his time in the city, I will not be giving any special status to his perspective. He will be very much a bit player in this story, reflecting his contemporary role as an undistinguished, and ultimately unsuccessful, medical student rather than the much grander part he was later to play in the broader history of evolutionary thought. As we will see, however, his testimony regarding his time in Edinburgh, in conjunction with the evidence from other sources, does raise some intriguing questions regarding the role of the Edinburgh transformists in the development of evolutionary thought in Britain later in the nineteenth century. It should not be surprising that many of the key evolutionary thinkers of the mid-nineteenth century, including Darwin himself, had been exposed to the heady intellectual atmosphere of Edinburgh in the 1820s. Out of the ferment of ideas that were current in Edinburgh in the 1820s and 1830s came the building blocks of the Darwin's evolutionary synthesis. While significant work has been done by Darwin scholars on transformism in Edinburgh during the narrow window of time from 1825 to 1827 when Darwin was resident in the city, much less is known about the years before and after these dates. Here I will for the first time be putting Darwin's Edinburgh years in the much wider perspective of the natural history scene in Edinburgh in the early nineteenth century.

In the two chapters that follow, I will flesh out the background to the story of transformism in Edinburgh, first exploring the University and medical education in the city in the early nineteenth century before turning the spotlight on Robert Jameson, his predecessor John Walker and the city's vibrant natural history circles. Having set the scene, I will then explore two central themes important to the reception of transformist ideas in Edinburgh: first, I will turn to the connections between transformism and the Wernerian doctrines in geology championed by Jameson and some of his closest associates; and, second, I will examine the close links between Edinburgh and Paris, at that time the most important international centre for the emerging biological sciences and the home of some of the key transformist thinkers of the period. In the penultimate chapter, I will explore the long-term significance of these developments and their possible impacts on the work of later transformist thinkers, including Robert Chambers and Charles Darwin.

## NOTES

1. Anderson, *Universities and Elites*, p.4.
2. See in particular Desmond, *Politics of Evolution*, Shapin, 'Phrenological knowledge' and Cooter, *Cultural Meaning of Popular Science*.
3. Shapin, *The Scientific Life*, p.1.
4. Shapin, *A Social History of Truth*, p.xxvii.
5. See, for example, Secord, 'Edinburgh Lamarckians' and Hodge, 'On Darwin's science and its contexts'.
6. The paper in which this suggestion has been most strongly made, Secord's 'Edinburgh Lamarckians: Robert Jameson and Robert E. Grant' (1991), first posed many of the questions that I will be addressing in this book.
7. Anon, 'Nature and importance of geology'.
8. Secord, 'Edinburgh Lamarckians', p.18.

# 2

# *Edinburgh's University and Medical Schools in the Early Nineteenth Century*

By the early decades of the nineteenth century the golden age of the Scottish Enlightenment was over. Most of the leading figures of that time were dead: the philosopher and historian David Hume had died in 1776, the philosopher Henry Home, Lord Kames, in 1782, Adam Smith the political economist in 1790, William Robertson the historian in 1793, James Hutton the geologist in 1797 and Joseph Black the chemist in 1799. The philosopher and historian Adam Ferguson lived on until 1816 and Dugald Stewart, the last significant disciple of the Scottish Common Sense school, was still professor of moral philosophy at the University of Edinburgh until 1820; but the days when Edinburgh had been one of the great intellectual powerhouses of Europe were passing. The high water mark of the reputation of Edinburgh's University too seemed to be in the past. In the mid-1820s Robert Mudie (1777–1842) observed that, with the exceptions of John Leslie (1766–1832), the professor of natural philosophy, Robert Jameson, the professor of natural history, and John Wilson (1820–51), the professor of moral philosophy,

> I did not hear that any of the Athenian professors have put in a single claim for immortality. Even in her anatomical school, that upon which she rested her fame the longest and the most securely, the recent falling off has been great; and in those who now shine in the list of her *senatus* there is none able to hold the book for Gregory, or the scalpel for old Monro, or light the furnace for Black.[1]

The University of Edinburgh, nonetheless, remained a force to be reckoned with internationally in education and scholarship. Its medical school in particular, although past its late eighteenth-century heyday, remained one the leading European centres of medical education and the most important and influential in the English-speaking world. The poor standard of medical training at the English universities and the religious qualifications they demanded for matriculation brought a steady stream of English students north, including many who were to play important roles in the development of the sciences in subsequent decades.[2]

University education in natural history was largely the preserve of medical men at this time. The subject, especially in its more speculative aspects, was dominated for most of the nineteenth century by men who had studied medicine at the University of Edinburgh, including Charles Darwin, Hewett Cottrell Watson (1804–81) and Richard Owen (1804–92). Natural history, which incorporated biology and botany, as well as geology, mineralogy and hydrography, was generally studied as part of a medical degree; science syllabuses leading to a degree were not to come into existence at Edinburgh until 1864. In this chapter I will explore how the University and its medical school in the early nineteenth century provided fertile ground for the reception and development of ideas that often conflicted sharply with the orthodox natural theology dominating its English counterparts.

## THE LEGACY OF THE SCOTTISH ENLIGHTENMENT

Many of the characteristics that made the University of Edinburgh strikingly different to Oxford and Cambridge, and distinctive even among the Scottish universities in the early nineteenth century, have their roots in developments that took place in the second half of the previous century. This was the period of the full flowering of the Scottish Enlightenment, when Scottish intellectual and cultural life was dominated by a remarkable group of men whose guiding values of tolerance, moderation and openness to new ideas had a profound influence on the University of Edinburgh. The result was a great efflorescence of learning in and around the University, and in particular at its famous medical school. In *The Expedition of Humphrey Clinker*, written in the late 1760s, the novelist Tobias Smollett puts the following words in the mouth of Matthew Bramble, the character whose views most clearly reflect those of the author himself:

> The university of Edinburgh is supplied with excellent professors in all the sciences, and the medical school, in particular, is famous all over Europe. – The students of this art have the best opportunity of learning it to perfection, in all its branches, as there are different courses for the *theory of medicine*, and the *practice of medicine*; for *anatomy, chemistry, botany*, and the *materia medica*, over and above those of *mathematics* and *experimental philosophy*; and all of these are given by men of distinguished talents.[3]

In the early decades of the eighteenth century the character of the Church of Scotland had little changed from the rigid and intolerant regime that in 1697 had presided over the trial and execution of an Edinburgh University student, Thomas Aikenhead, for blasphemy, the last execution for this crime in British history. However, by the middle decades of the century a new spirit had come to dominate the church, one associated with a group of highly cultured churchmen who have been fittingly described as the 'Moderate literati of Edinburgh'.[4] As Smollett wrote, 'the kirk of Scotland, so long reproached with fanaticism and canting, abounds at present with ministers celebrated for their learning, and respectable for their moderation'.[5] The rise of the Moderate literati owed much to the patronage of the 3rd Duke of Argyll, Lord Islay, who was Prime Minister Robert Walpole's political manager in Scotland and has been described as the country's 'chief patron' from around 1723 until his death in 1761.[6] He shared the tolerant, broadly secular outlook of his Moderate protégées and had little time for the Popular Party in the Kirk, whose views harked back to the inflexible Calvinism of the late seventeenth century. This group of cultured and tolerant but socially conservative men came to dominate the general assembly of the Church of Scotland and established a similar ascendency over the Scottish universities, and in particular the University of Edinburgh. The core of this group were William Robertson, Adam Ferguson, Hugh Blair, John Home and Alexander Carlyle. Apart from Carlyle, all these men were professors at the University of Edinburgh, and Robertson was its principal during a crucial period in the late eighteenth century from 1762 until 1793. Thanks to them, in the words of Sher, '[i]n no other country were the principles of the Enlightenment so deeply rooted in the universities'.[7]

The conflict within the Kirk over the works of their friends David Hume and Lord Kames in the mid-1750s gave an early opportunity for the Moderate literati to defend their own ideals in a showdown with their opponents of the Popular Party. The controversial writings of Hume and Kames had been condemned as inimical to true religion by many of the ministers of the Popular Party, who saw themselves as the guardians of Calvinist orthodoxy. The Moderates successfully defended their friends, arguing that, while actions harmful to society should be punished, the expression of erroneous ideas, whether religious or secular, should be tolerated. However, their espousal of freedom of thought and expression did not incline them towards political radicalism. Their social and political views were in fact staunchly conservative and they upheld the Hanoverian political and social status quo at least as strongly as they stood for intellectual freedom and religious toleration.

Sher has described the election of William Robertson as principal of the University of Edinburgh in 1762 as 'probably the most important single event of the eighteenth century' for 'the institutionalization of Moderate authority and Enlightenment values in Scotland'.[8] Moderate dominance of the University was to have profound consequences for the nature and quality of the education it offered. Many of the figures who added lustre to the teaching of the sciences and medicine at the University in the late eighteenth century, such as Joseph Black (1728–99), John Robison (1739–1805), John Playfair (1748–1819) and William Cullen (1710–90), were friends, or sometimes relatives, of the group of Moderates around William Robertson, who dominated the Church and the University in this period, and owed their chairs to Moderate patronage. John Walker was also a Moderate clergyman before being appointed to the chair of natural history, which he held until his death in 1803. Despite owing their positions to connections with Moderate patronage networks, many of these individuals turned out to be exceptional teachers and original thinkers in their respective fields. It is largely their influence that, in the words of J. B. Morrell, gained for Edinburgh in the late eighteenth century 'the reputation of being the best university for science in Europe and in the English-speaking world'.[9]

Robertson's legacy was to last well into the following century. As Dugald Stewart, who held the chair of moral philosophy at Edinburgh between 1785 and 1820, was to write of Robertson, 'if, as a seat of learning, Edinburgh has, of late more than formerly, attracted the notice of the world, much must be ascribed to the influence of his example, and to the lustre of his name'.[10] In particular, Stewart praised Robertson's role in encouraging the establishment of the literary and medical societies that 'contributed so essentially to the improvement of the rising generation'. The tradition of Edinburgh student societies, where daring ideas could be discussed without fear among like-minded peers in an atmosphere of tolerance and free enquiry, played an essential role in the cultural life of the University. While most of the first generation of Moderate literati were dead by the beginning of the nineteenth century, the University was still very much dominated by their friends and disciples in the first few decades of the new century. The values of freedom of thought and expression and religious toleration still permeated the institution and created a very different atmosphere to that found in the intellectually conservative English universities. These were still dominated by the Anglican clergy in this period and would hold a monopoly on university education in England until the foundation of the secular University College London in 1826.

## THE UNIVERSITY OF EDINBURGH AT THE BEGINNING OF THE NINETEENTH CENTURY

At the beginning of the nineteenth century Scotland had five universities: Edinburgh, Glasgow, St Andrews, and King's College and Marischal

College in Aberdeen. In the late eighteenth century Edinburgh had been predominant among the Scottish universities in science and medicine. However, in the early decades of the nineteenth century it was losing its pre-eminence. The medical school of the University of Glasgow was rapidly coming to rival Edinburgh, and south of the border Cambridge saw a revival of research in botany and geology. The foundation of University College London in 1826 also provided Edinburgh with a non-denominational rival in London.[11] In the mid-1820s the incumbents of only three of Edinburgh's thirty-two professorial chairs could claim an international reputation: Robert Jameson, professor of natural history; John Leslie, professor of natural philosophy; and Thomas Charles Hope (1766–1844), professor of chemistry and medicine. Nonetheless, the University continued to attract students from across the British Empire and beyond.

We are lucky to have an important source of information on the state of the Scottish universities in the mid-1820s in the form of the report of the Scottish Universities Commission of 1826, established by the then home secretary Robert Peel in response to growing criticism of the operation and administration of the universities. Unusually, the University of Edinburgh had been founded by the town council as the City College, and the council still played a major role in the running of the University and the appointment of professors. Conflicts between the town council and the authorities of the University of Edinburgh over such issues as patronage and the appointment of new professors played a major role in the decision of Peel to establish the Scottish Universities Commission with the aim of recommending reforms. The findings of this report were published in 1830. The section of the report relating to the University of Edinburgh and the detailed testimonies gathered there from the professors and other interested parties paint a vivid picture of the state of the University in the second half of the 1820s. The report clearly reveals that 'the importance of personal connexions and personal whims, as opposed to qualifications and rules of procedure, produced unexpected and unpredictable behaviour; professors were at least as loyal to their clients as to the University; and the location and extent of various sorts of authority were not sharply defined'.[12]

One of the specific issues that was causing concern was the open conflict that had broken out between Edinburgh town council, who were responsible for the appointment of two-thirds of the chairs, and the Senate of the University, composed of the University's principal and professors. This dispute centred on the proposal made in 1824 that the course of James Hamilton (1767–1839), professor of midwifery, be made compulsory for graduation in medicine. Hamilton was supported in this by the town council but opposed by the Senate, who saw this as interference with their authority over requirements for graduation. The Senate itself had petitioned Peel to help arbitrate in this dispute.

The University was composed of four faculties: Arts, Divinity, Law and Medicine. Under the traditional Scottish university system students normally enrolled at the age of 16, but sometimes at fifteen or even younger, to study for four years in the Arts faculty. Those wishing to graduate with a Master of Arts studied Latin, Greek, rhetoric, mathematics, natural philosophy, logic and moral philosophy. During the eighteenth century there was a great decline in the number of students taking the Master of Arts degree at Edinburgh. In his history of the University, Alexander Grant suggested that numbers had fallen from a peak of 104 in 1705 to only one or two a year towards the end of the century.[13] Despite periodic attempts by the University to encourage students to graduate, the number of students graduating with an MA remained extremely low into the early decades of the nineteenth century. There was little incentive for students to graduate from the Arts faculty, as this was not a requirement to study for the higher degrees of law, medicine or divinity. The choice of courses pursued by students was therefore determined only by their own inclinations, their ability to pay class fees and their assessment of what would prove useful preparation for any further study they required if they intended to gain entry to one of the professions. As Grant wrote, 'each student attended such classes as he or his friends might think advisable'.[14] With the important exception of the Medical faculty, few students followed the full degree programme. There was also no restriction on who could attend a course except a student's ability to pay. This lack of admission requirements made sense from the professors' perspective as they had a strong pecuniary interest in maximising the numbers of students taking their classes. In subjects where the content of the lectures was of interest to particular professional groups, or attracted a large number of amateur enthusiasts, as was the case with chemistry and natural history, a large proportion of the class were often non-graduating students.

Students could commence their studies of Divinity, Law or Medicine at the age of nineteen or twenty to prepare them for one of the professions. Preparing young men for the professions was a major function of the University of Edinburgh. Oxford and Cambridge, by contrast, had largely given up vocational education, except for supplying clergy to the Church of England. For the sons of the English aristocracy and gentry they provided, in the words of R. D. Anderson, 'a social finishing school as much as an intellectual experience'.[15] This situation only began to be remedied with the founding of University College London in 1826. Medical students at Edinburgh were required to study for three years, which was raised to four years in 1825. However, if students had taken a Master of Arts degree or gained previous medical experience working in a hospital or with the army, navy or East India Company, they were required to study for only three years. Study at another university was also taken into account, and it was very common for students from outside Scotland to come to Edinburgh to study for a final year to round off their medical education before graduating. The high proportion of students from outside Scotland marked out the Medical faculty, as Scotland's national Church and unique legal system made study in these faculties less attractive to students from other parts of the United Kingdom or further afield. Unlike Oxford and Cambridge, Edinburgh did not attract any significant proportion of the sons of the English and Irish aristocracy, who made up an 'infinitesimal' proportion of the student body.[16] The vast majority of English students who studied at Edinburgh came from middle-class backgrounds. Very many of them were from Non-conformist families, attracted by the absence of any religious test for entry to the University of the kind that existed at the Anglican strongholds of Oxford and Cambridge.

That students at the University of Edinburgh came from a wide variety of social backgrounds is attested by the report of the Scottish University Commission, who, while acknowledging that some students came from wealthy backgrounds, recognised that 'there are also many in very straitened circumstances'.[17] The relatively low class fees (£4 4s for the natural history course in 1822[18]), the absence of residential requirements, the wide availability of bursaries for poorer students and an ethos that emphasised hard work and academic achievement over 'aristocratic dissipation' were some of the factors that made the Scottish universities in general much more socially inclusive than their English counterparts.[19] Poorer students were often obliged for financial reasons to leave the University a considerable time before the end of the session. This was such a problem that many professors took to granting class certificates on 1 April so as not to further disadvantage these students. The commission were sensitive to these issues and opposed any lengthening of the session that would further exacerbate the problem and would be liable to 'cut [poorer students] off from University education altogether', remarking that '[i]t would be a strong measure to adopt any regulation which would suddenly produce such effects as these'.[20] The relatively democratic atmosphere, especially when compared to Oxford and Cambridge, is attested to by J. G. Lockhardt, the son-in-law of Sir Walter Scott, who wrote disapprovingly of Edinburgh students in a work published in 1819 that 'the greater part of the company are persons whose situations in life, had they been born in England, must have left them no chance of being able to share the advantages of our academical education'.[21] The Edinburgh student body was therefore much more socially mixed than at the English universities of the period and provided students with opportunities to become acquainted with peers from a variety of different social backgrounds.

Tutorials had traditionally formed an important element of teaching for almost all courses at Scottish universities, although there is some evidence that this practice had fallen into abeyance on the part of some of the Edinburgh professors in the early nineteenth century, in particular those of logic, moral philosophy and natural philosophy.[22] These tutorials generally took the form of 'examinations', in which the professor would pose questions that the students were expected to answer and discuss. This gave students the opportunity to debate the topics presented to them by their professors in the lectures rather than just be passive recipients of the wisdom handed down to them. Students would be set essays and encouraged to present their own work and critique that of others in the presence of the professor and the rest of the class. In the eighteenth century this teaching method had been applied with great effect by William Cullen, professor of chemistry and medicine at Edinburgh. In his inaugural address on his accession to the chair of chemistry in 1858, Lyon Playfair gave a laudatory account of Cullen's teaching style, noting how he 'saw that a science like Chemistry was not to be taught by mere lectures, but that there must be a free and unreserved communication between the teacher and his pupils'.[23] Robert Jameson also appears to have been a model professor in this regard. He not only held discussions with students for an hour before his lectures, but also met with them for more informal conversations between three and six times a week in the University's natural history museum.[24] Most of the professors, with only a few exceptions, followed the example of George Jardine (1742–1827), Glasgow's professor of logic and rhetoric from 1787 until 1824, who had championed this approach to teaching in his *Outlines of Philosophical Education* (1818 and 1825). In this work he strongly advocated supplementing lectures with 'a system of active discipline on the part of his students, with a view to invigorate, and improve, the important habits of enquiry and of communication'.[25] The Scottish Universities Commission Report of 1826 praised Jardine's system highly and commended its application by most of the Edinburgh professors, commenting that 'when examination is properly conducted, it becomes even fascinating to the youthful mind – and that it is so essential, that there is at least the same reason for enforcing it as part of the business of the class, as there is for enforcing the Lectures to which it should be made to relate'.[26]

The professors at the Scottish universities were primarily concerned with teaching rather than research. Any research that they did conduct was additional to their teaching duties. A chair came with the obligation to lecture and in return the professors had the right to charge students fees for attending the course, which the professors were responsible for collecting at the start of each session. In contrast with the 'endowment' system prevalent at Oxford and Cambridge, the bulk of Scottish professors' incomes came directly from their students. Their remuneration therefore depended to a considerable extent on the size of the class they could attract. In addition to the student fees, most professors also received a modest salary, although this was not the case for a substantial proportion of the Medical faculty. The professors of medicine, however, were at the considerable advantage of being able to supplement their earnings through medical practice. Chairs established by the Crown came with a modest salary of between £30 and £300 pounds in 1826, while those nominated by the town council received between £22 3s 4d and £196 2s 2d, but this was generally insufficient to maintain the lifestyle appropriate to the social status of a university professor.[27] This was particularly true at Edinburgh because it was controlled by the town council. Because of this the university therefore lacked fiscal autonomy and was not in a position to pay the professors more ample salaries. If a professor did not have independent means, he therefore had to take his teaching responsibilities extremely seriously. Eight of the professors, all belonging to the medical school, received no salary of any kind from either source, although four of them succeeded in making enough in student fees alone to put them among the eight highest earning professors at the university. Nonetheless, a popular lecturer stood to make a very comfortable living from his chair. Edinburgh's professor of chemistry, Thomas Charles Hope, who attracted large crowds of Edinburgh's citizens to see his spectacular demonstrations, earned £2,213 8s in 1826, all of it from student fees, as he received no salary from either the Crown or the town council.[28] In the spring of that year he introduced a new popular course in chemistry, replete with spectacular experiments and demonstrations. This course also admitted women, and proved immensely popular with the middle-class ladies of Edinburgh. His success attracted both envy and scorn in some quarters. Henry Cockburn (1779–1854), the Edinburgh lawyer judge and literary figure, for example, wrote in a letter to his friend Thomas Francis Kennedy in February 1826: 'I wish some of his experiments would blow him up. Each female student would get a bit of him.'[29]

Under such a system, the professors were in effect freelance teachers who had been licensed to teach by the town council and the Senate of the University. The 1826 Universities Commission proposed strengthening the regulatory role of the university authorities and curbing to a certain extent the autonomy of the professors. To this end they proposed 'that the

University-Court should be invested with general powers of superintendence, which will be sufficient to prevent or correct any accidental failure in the regular discharge of the duties of the Professors'.[30] However, as we will see below, these proposals for reform came to nothing. While the existing system could be, and was, criticised, it actually tended to promote teaching that met students' needs, as it gave professors a strong interest in maintaining the quality of their courses. As the professors depended on the course fees for their livelihoods, the professorial chairs were unlikely to degenerate into sinecures, as often occurred at the English universities, where chairs were often handsomely endowed. This was perhaps more true for those professors at Edinburgh whose courses were not a requirement for graduation, such as natural history, than for those such as anatomy, which were obligatory for students wanting to graduate with an MD. The effects of market forces on teaching were not, however, always positive. For example, John Leslie, professor of natural philosophy from 1819 until his death in 1832, found he had to lower the level of his lectures to accommodate those students with limited mathematical abilities or risk losing money.[31]

While this method of remuneration may have encouraged professors to devote more energy to their teaching, it could also provide a positive disincentive to engage in research. Hope, the professor of chemistry, whose earnings from course fees dwarfed those of other professors, largely abandoned research and ceased publishing. It was presumably with such examples in mind that the natural philosopher and science writer David Brewster (1781–1868), later to become principal of the University of Edinburgh, was to write in a review of Charles Babbage's *Reflections on the Decline of Science in England* in 1830 that the university professor 'is forced to become a commercial speculator, and under the dead weight of its degrading influence, his original researches are either neglected or abandoned'.[32]

Although professors were reliant on income from the course fees, they did have the benefit of life tenure. According to Grant, the principle that professorial chairs were held for the lifetime of the incumbent was established on the precedent of the terms offered to Alexander Monro, primus, in the early eighteenth century, and subsequently extended to the other professors. Grant considered that this greatly augmented the professors' 'independence and respectability'.[33] While it was true that this gave professors a measure of security, it also meant that incumbents often hung on to their chairs long after age or incapacity had made them incapable of meeting their teaching obligations. In such cases the professor often paid a substitute to do the teaching for him. Although most Edinburgh chairs were in the gift of the Crown or the town council, who also had to approve the appointment of a substitute, this system provided a route whereby professors could effectively choose and groom their successors, not infrequently their own sons or other relatives. Nepotism was therefore rife. The Monro dynasty that monopolised the chairs of anatomy and medicine at Edinburgh for three generations from 1720 until 1846 was a particularly egregious example. Unsurprisingly, given the importance of patronage and personal connections in securing a chair in this period, the vast majority of professors across all the Scottish universities were Scots. In the period 1800–19 all professors appointed across the Scottish universities were born in Scotland, although the figure for 1820–29 had dropped to 88 per cent.[34]

The Scottish Universities Commission of 1826 was sharply critical of the practice of appointing joint-professors, or so-called 'successors', while the existing professors continued in their chairs. They declared this was 'a practice which, after careful consideration, We are of the opinion ought in general to be prohibited, as giving facilities to the appointment of persons who in all probability might not have been selected, if an opportunity had been given by a declared vacancy for a free competition for the office'.[35] The commissioners, however, recognised that if professors were unable to pass some of their commitments on to assistants when they became too old or too ill to teach while still continuing to hold the chair, alternative provision would need to be found for their support. Although the Commission of 1826 was highly critical of the practices of the University in many ways, it did not result in any substantial change. This would have to wait until the Universities Act of 1858, which brought in sweeping reforms.

Although some professors came from more modest backgrounds, they were generally recruited from among the middle classes. As R. D. Anderson notes:

> they came above all from two groups, the prosperous professional and business classes, and a more modest bourgeois stratum of ministers, farmers, schoolmasters, and small businessmen; below these was a smaller group from the fringe of the middle and working classes, the sons of shopkeepers and skilled workers.[36]

Whether the appointment of a particular professor was made by the Crown or the town council, political influence and patronage were crucial in determining the successful candidate. To stand a chance of success it was therefore essential for a candidate to be plugged in to local and national patronage networks. With the right connections, it was nonetheless possible for talented candidates from relatively modest backgrounds to gain university chairs. John Leslie, for example, was the son of a joiner, but still managed to become professor of mathematics at Edinburgh from 1805 before moving to the chair of natural philosophy in 1819.[37] His candidacy had been championed by leading figures within the Evangelical Party of the Church of Scotland, including a young David Brewster, who wrote an anonymous pamphlet in support of his candidacy under the pseudonym 'A Calm Observer'.[38] Leslie's candidacy became something of a *cause célèbre* after it was opposed by some within the Moderate Party of the Church on the grounds that he had written approvingly of the notoriously sceptical philosopher David Hume in an endnote to his work *An Experimental Inquiry into the Nature and Propagation of Heat*, published the previous year. In this book, Leslie had written, 'Mr Hume is the first, as far as I know, who has treated causation in a truly philosophic manner. His *Essay on Necessary Connexion* seems a model of clear and accurate reasoning.'[39] This was enough for some Moderates, eschewing the tolerance so characteristic of the Moderate literati of the previous century, to consider his character questionable from a religious standpoint. His candidacy became a trial of strength between the Moderate and Evangelical Parties within the church, and his appointment a resounding defeat for the former and their chosen candidate the Reverend Thomas MacKnight. It might seem surprising at first sight to observe Evangelicals supporting an admirer of the arch-sceptic Hume, while Moderates strenuously opposed him, especially when it is remembered that many of the Moderate leaders of the previous century were personal friends of Hume and had defended him in the General Assembly of the Kirk. However, it should be borne in mind that Leslie's Evangelical champions rooted their religious convictions in personal faith and put little store by natural theology founded on the argument from design so effectively demolished by Hume. In any case, Moderate opposition to Leslie made it politically expedient for the Evangelicals to throw their weight behind him.

Morrell has commented on the diffuse nature of authority within the University in the 1820s, describing its administration as 'pre-bureaucratic'.[40] Power relations within the University were mediated more by a network of patron–client relationships that extended well beyond the University itself than by regular administrative structures within it. With the gift of most of the chairs in its power and overall responsibility for the supervision of the University, the town council had enormous influence. Some sense of how the council wielded this influence in the 1790s can be gathered from Henry Cockburn's description of them in his *Memorials of his Time*: 'Silent, powerful, submissive, mysterious, and irresponsible, they might have been sitting in Venice.'[41] Despite this, the town council, composed as it was of merchants and businessmen with limited knowledge of medicine, had been notably successful in appointing talented professors during the eighteenth century, due in no small part to the influence of the dominant Moderate Party within the Church and University. This had done much to establish and maintain the reputation of the university as an international centre of excellence, especially in medical education. However, by the early nineteenth century their choices were not always so happy. In Mudie's opinion the system of patronage, fatally combined with 'civil ignorance, political influence, and clerical intrigue' had done lasting damage to the University.[42] Patronage directed by the enlightened Moderatism of Robertson had given way to that of an ossified and increasingly conservative Moderate establishment by the early nineteenth century. Jacyna may well be correct when he traces this hardening of political attitudes to a conservative reaction of the Edinburgh establishment to the French Revolution and its Scottish sympathisers.[43] This view certainly accords with Cockburn's description of Edinburgh in the early decades of the nineteenth century, where '[e]verything rung, and was connected, with the Revolution in France; which for above 20 years, was, or was made, the all in all. Everything, not this or that thing, was soaked in this one event'.[44] In the early decades of the new century the Tory-dominated council often made blatantly political appointments, as when they installed James Home, previously professor of materia medica, as professor of the practice of medicine (also sometimes referred to as practice of physic) in 1820, over the superior claims of the Whig John Thomson, professor of military surgery.[45] Once appointed, the professors were often a law unto themselves. They enjoyed a legal monopoly on the teaching of their subject and had absolute authority over what was taught and how. While this high level of autonomy had been a positive benefit to the likes of Cullen and Black and their students, in the hands of less talented or committed professors this level of freedom could have less happy consequences for teaching quality.

Despite talk of decline, the University of Edinburgh actually reached a peak in student numbers in the 1820s, attracting 2,236 students in 1825,[46] but this had dropped to 1,056 by 1844 (although numbers picked up again later in the century).[47] This was a pattern common to all the Scottish universities, although it was particularly pronounced at Edinburgh. Of the students enrolled at the University in 1825, 892 of them were studying medicine, making this by a significant margin the most popular course of study. The large number of students at Edinburgh caused problems for teaching and assessment, identified in the report of the Universities Commission of 1826. The commissioners wrote that:

> Examinations are not attempted by a number of the Professors, and when they are, from the number of Students, and from the limited time, they are so

carried on as not to afford that stimulus to exertion, which, when efficient, they supply. A separate hour is not taken for the Examinations as part of the regular business of the class; and it is left in many classes to the Students themselves to undergo the Examinations or not as they please. Essays are prescribed, but the performance of them is not compulsory; and it is in evidence, that even the countenances of those who attend some of the classes, are scarcely known by the Professor, and the attainments of a great number not at all.[48]

The intellectual life of the University was greatly invigorated by the activities of the many student societies. These included the Royal Medical Society, founded in 1737, and the Royal Physical Society, founded as the Physico-Chirurgical Society in 1771. Both of these continued in existence throughout the nineteenth century and beyond. Such societies provided fora for students to discuss all manner of theories and ideas in a more informal setting and without the presence of their professors. Although the Royal Medical Society was specifically for medical students, it did not confine itself to medical matters; papers given at its meetings ranged across all manner of scientific and philosophical topics. Many years later Henry Brougham remembered addressing the Society on the subject of 'liberty and necessity' while he was a student at the University in the 1790s, an experience he describes as greatly stimulating his desire to master the art of oratory.[49] Jacyna has commented on the 'tendency of student medical societies to become cockpits' where radical and unorthodox ideas could be freely aired and debated.[50] Writing of his own time as a student at the University in the mid-1780s, Sir James Macintosh later recalled: 'Every mind was in a state of fermentation. The direction of mental activity will not indeed be universally approved. It certainly was very much, although not exclusively, pointed towards metaphysical enquiries.'[51] It is noteworthy that Mackintosh went on to add, not without a note of disapproval: 'Speculators could not remain submissive learners.' As we will see in the next chapter, some of these student societies were still playing an important role in the early decades of the nineteenth century in promoting speculation and debate on the hot scientific and philosophical topics of the day.

In addition to the attractiveness of some of the teaching and the long-standing reputation of the University, another factor that drew many English students to Edinburgh was the absence of any religious restrictions at the Scottish universities of the type in force at their English equivalents, which demanded that students submit to the Anglican Confession of Faith as a requirement for matriculation in the years before the establishment of the secular University College London. While Oxford and Cambridge held to the principle that 'religious and secular education were inseparable' and clung to the 'persisting ideal of a national elite with common values and experiences, which secularization would destroy', Edinburgh had long been a largely secular institution.[52] This made it particularly attractive to students from Non-conformist backgrounds, who could not in good conscience subscribe to the Thirty-Nine Articles of the Church of England. Alexander Bower published a guide for prospective students at the University in 1822 in which he noted that 'no oath, nor subscription to any article of religion, nor Confession of Faith, are required, as is the case at the universities of Oxford and Cambridge. Persons of every profession of religion are freely admitted, whether Catholics or Protestants, and no questions are asked'.[53] This open and tolerant attitude in matters of personal conscience doubtless made Edinburgh especially attractive to students for whom the clerical atmosphere of the English Universities would have been uncomfortable, if not downright hostile. In Edinburgh students could achieve a considerable degree of both social and intellectual independence, a far cry from the paternalistic atmosphere of Oxford and Cambridge's residential colleges.[54]

There also appears to have been as little attempt to guard against heterodox views among the professors as among the students. Unlike at Oxford and Cambridge, the professors were generally not churchmen, with the exception of the Principal and, until his death in 1803, John Walker, the professor of natural history, who was a minister of the Church of Scotland. As the 1826 Universities Commission noted disapprovingly, '[i]t has not been the uniform practice of Professors admitted into the University of Edinburgh, to produce, in terms of the act of Parliament 1707, and act 14th of the General Assembly 1711, at their admission, any certificate of their having subscribed the Confession of Faith, or taken the oath of allegiance.'[55] The commissioners went on to add that when George Baird, the principal of the University from 1793 to 1840, was asked if the oath of allegiance was effectively 'inoperative' at Edinburgh he simply answered 'Yes'. The practice of administering a religious test to newly appointed professors had in fact fallen into abeyance while William Robertson was principal at the end of the previous century, reflecting the tolerant attitudes prevalent in the Church establishment in this period.[56] It seems reasonable to suggest that this was the legacy of the 'Moderate literati', who had dominated the Church and University in the late eighteenth century and who tolerated alternative religious and philosophical opinions to an extent that sometimes earned them the ire of more inflexible churchmen.[57]

The type of education available at the Scottish Universities in the early nineteenth century was clearly fundamentally different from that available at the two English universities at that time. The nature and import of these differences has been the subject of a lively debate sparked by the publication of George Davie's *The Democratic Intellect* in 1961. In this immensely influential work Davie argues that 'by 1830, a severe crisis had arisen in Scotland on the question of how far the Universities were to subordinate themselves to the Southern system'.[58] Davies sees the nature of the Scottish Universities Commission report of 1826 to be essentially 'a surprise attack on the national academic tradition, delivered by a group of influential Scots who wished to impose Southern standards'.[59]

To support his argument that the commissioners 'rather regarded the English University tradition as the proper model to follow' he cited the evidence presented to the Commission by Archdeacon Williams, rector of Edinburgh Academy.[60] Williams had suggested the replacement of the traditional philosophy-dominated curriculum with one based on the study of Greek. Davie also noted that Sir Daniel Sandford, professor of Greek at the University of Glasgow, made similar recommendations to the Commission. However, while the Commissioners did advocate raising the standard of teaching in classical languages at the Scottish universities, in general their conclusions did not altogether support Davie's contention that Williams' and Sandford's views triumphed. In the *Report Relative to the University of Edinburgh* the Commissioners in fact made an explicit plea in defence of Scottish particularism against those who might want to bring the Scottish universities into line with Oxford and Cambridge. The Commissioners argued that, while graduates of Oxford and Cambridge might aspire to high positions in the Anglican church hierarchy or to one of the endowed chairs at the English universities themselves, such opportunities did not exist in Scotland:

> Even to those who are destined for the learned professions, there is little motive for paying almost exclusive attention to the ancient languages, or for seeking to acquire that profound acquaintance with all the niceties and difficulties of the Latin and Greek, which is in England venerated as the undoubted evidence of learning. Our Ecclesiastical establishment affords no situations which are to be attained by such an acquisition, or in which there is leisure for increasing it.[61]

The Commissioners went on to argue that the Scottish universities were well adapted to their function, which was significantly different to that of Oxford and Cambridge. An education that concentrated on the classical languages or mathematics, as at the English universities, at the expense of the broader philosophy-based education offered in Scotland would not meet the needs of Scottish students, or those from Dissenting backgrounds from other parts of Great Britain. They foresaw that the imposition of the English model on the Scottish universities and their curricula would have dire consequences for their future:

Were such a measure to be adopted, the unavoidable consequence would be, that all who have no intention of being Literary men, and that all the Dissenters, who form a prodigious proportion of the youth attending our Philosophy Colleges, would be banished, virtually at least, from them, and that even Edinburgh would, except for Medicine and Law, be comparatively deserted, while the less frequented Universities would lose nearly two-thirds of their Students.[62]

However the recommendations of the Commission are interpreted, one thing is certain; they did not lead to any significant reform of the Scottish universities. A bill brought before Parliament in 1836 to appoint a board of visitors to carry out the recommendations of the Commission was quietly dropped after vociferous opposition in Scotland to any such interference in the university system. The recommendations of the report were therefore never implemented and there was no realistic threat to the survival, at least for the time being, of what Henry Brougham, a graduate of the University of Edinburgh himself, described as 'a system which cultivated and cherished higher objects than mere learning, which inculcated a nobler ambition than the mere acquisition of prosody and dead languages'.[63]

## THE UNIVERSITY OF EDINBURGH'S MEDICAL SCHOOL

Although past its late eighteenth-century apogee, Edinburgh still had the most highly regarded medical school in the English-speaking world in the early nineteenth century and could boast of a worldwide reputation. The classes of the medical professors indeed continued to grow in size into the early decades of the century. MD graduations showed a trend diametrically opposite to the decline in the number of MA graduates. In the first half of the eighteenth century only half a dozen or so students graduated with a medical degree every year. The numbers graduating increased steadily throughout the second half of the eighteenth century and the early decades of the nineteenth. In 1824, 140 students took the MD degree, while the high water mark of 150 students was reached in 1827, the year Darwin left Edinburgh. This number was not to be exceeded until after the reforms brought about by the Universities Act of 1858.[64] In its heyday in the late eighteenth and early nineteenth centuries Edinburgh produced more than half of Britain's systematically trained doctors and a third of those serving with the army, navy and East India Company.[65]

A significant proportion of the students were from outside Scotland, the vast majority of these being English. As education in natural history in the early nineteenth century was largely the preserve of medical students, it is therefore no surprise to find that many of the most important figures in nineteenth-century British biology had studied medicine at Edinburgh.

Among many others, these included the comparative anatomist Richard Owen, the physiologist William Benjamin Carpenter (1813–85), the physical anthropologist James Cowles Prichard (1786–1848), the botanist Hewett Cottrell Watson and, of course, Charles Darwin himself.

In 1826 the Medical faculty of the University had eleven professors: anatomy and medicine (Alexander Monro, tertius), chemistry and chemical pharmacy (Thomas Charles Hope), botany (Robert Graham), materia medica (Andrew Duncan junior), theory of physic (William Allison), practice of physic (James Home), midwifery (James Hamilton junior), pathology (John Thomson) clinical medicine (held jointly by 'the four medical professors'), clinical surgery (James Russell) military surgery (Sir George Ballingall) and medical jurisprudence (Robert Christison). The chairs of clinical surgery and military surgery had been relatively recently created by the Crown to meet the needs of the armed forces for military surgeons during the wars with France, against the opposition of the existing medical professors. The University Commission Report also noted the existence of a

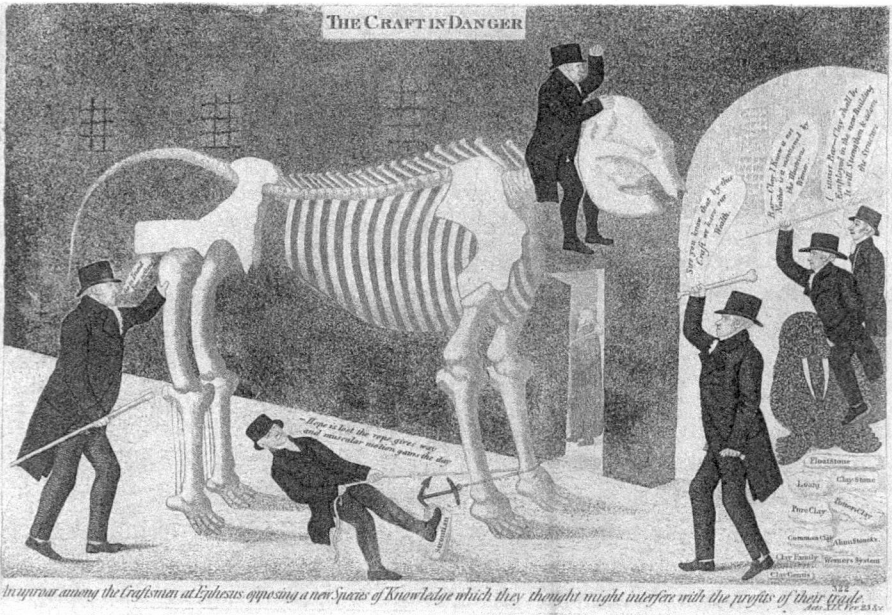

Figure 2.1 John Barclay attempts to enter the University in 1817 astride the skeleton of an elephant and opposed by professors John Hope, Robert Jameson and Alexander Monro, tertius. Barclay was considered the prime candidate for a proposed new chair of comparative anatomy. This proposal was defeated by fierce opposition from the established professors. From John Kay, *A Series of Original Portraits and Caricature Etchings*, vol.2, part 2, p.448. Credit: Wellcome Collection. CC BY.

chair in surgery and stated that it was currently vacant, although elsewhere in the *Report Relative to the University of Edinburgh* it is suggested that surgery was in fact also taught by Monro at this time. Incumbent professors guarded their prerogatives and the succession of their chairs fiercely. For example, in 1817 a conflict broke out over the proposal to found a chair of comparative anatomy at Edinburgh, for which the extra-mural anatomy lecturer John Barclay (1758–1826) was the prime candidate. While a few of the professors supported him, the proposal was defeated by the concerted opposition of a cabal of the established professors, led by John Hope (at that time professor of botany), Robert Jameson and Monro, who felt it would impinge on their territory. In the end the proposal was abandoned.

At least part of the blame for the decline of the reputation of the Edinburgh medical school in the early decades of the nineteenth century must be laid at the door of Monro. He was the third generation of his family to hold the chair of anatomy at Edinburgh. His grandfather, Alexander Monro, primus (1687–1767), had been an excellent and innovative professor, who had studied at the University of Leiden under the great Herman Boerhaave (1668–1738). His son, Alexander Monro, secundus, who succeeded him, also proved to be a highly competent professor of anatomy and medicine from 1785 to 1817. The first Alexander Monro's grandson, however, did not prove to be cast from the same mould as his father and grandfather. Darwin, never one to mince his words when criticising professors he did not respect, did not spare Monro. While a student at Edinburgh he wrote to his sister Caroline of the professor of anatomy: 'I dislike him & his lectures so much that I cannot speak with decency about them. He is so dirty in person & actions.'[66] Decades later he recalled in his autobiography how 'Monro made his lectures as dull, as he was himself'.[67] It was even reported that Monro read his grandfather's lecture notes verbatim to his class, not even taking the trouble to excise references such as 'When I was a student in Leiden in 1719 . . .'[68] Although the authenticity of this account is questionable, the fact that this story was in contemporary circulation gives a strong sense of how he was regarded by his students and peers. Attendance at his classes reflected the poor opinion of his lectures on the part of the medical students; in 1799 306 out of 417 medical students attended his lectures, while by 1821 only 205 out of 817 troubled themselves to attend. This compares very unfavourably with attendance in the time of Monro, tertius' father, whose class was attended by 333 out of 387 students in 1797–8.[69] The poor quality of lecturing in anatomy and surgery meant that those students who took their studies seriously often felt they had to turn to the large number of extra-mural teachers to supplement the teaching within the University.

The Monros did not have a monopoly on nepotism within the faculty of medicine; between 1786 and 1807, of the ten appointments made to medical chairs, eight were also sons of Edinburgh professors. However, not all of the medical professors were as lacklustre in their teaching and complacent in their attitude as Monro, tertius. James Home, professor of the practice of physic from 1820, for example, introduced examinations into his teaching, which had not been the practice of his predecessor James Gregory. Home himself advocated this practice before the Scottish Universities Commission as not only a valuable stimulus to diligent study on the part of the students, but also as good way for him to become better acquainted with them and get a better appreciation of their individual qualities and capacities.[70] Medical students at Edinburgh could also benefit from extremely valuable first-hand experience of clinical medicine and surgery not available elsewhere. As Chitnis has noted, '[c]linical medicine and surgery were introduced into Britain at Edinburgh and by 1826 it was still almost confined to the Scottish capital'.[71] Access to the wards of the Edinburgh Royal Infirmary provided medical students with invaluable hands-on experience not available elsewhere.

In order to graduate with the degree of MD, students had to produce certificates to prove they had attended the following courses: anatomy, chemistry, materia medica, theory and practice of medicine, midwifery, clinical medicine and botany. They also had to show they had attended two from among practical anatomy, natural history, medical jurisprudence, clinical surgery and military surgery.[72] In addition, they were required to write and have printed in Latin a thesis on a medical subject. In 1833 the rules were relaxed to allow students to present their theses in either English or Latin, and from 1835 practically all were written in English. Students had to defend this thesis in an oral examination conducted by two professors. They would also be examined by two professors on the various branches of medicine in the presence of the faculty. In addition, they would have to produce written commentaries on two aphorisms of Hippocrates and two case reports, which they would then also have to defend before the faculty. All of these exercises, both written and oral, were conducted in Latin until 1833. Once they had successfully undergone these trials, students were eligible to graduate with the coveted MD from the University of Edinburgh.

Although the subjects students had to take in order to graduate were clearly prescribed, there was no overarching course of study that students had to follow, and they were free to take the classes when and in whatever order they chose. As the Universities Commission pointed out disapprovingly: 'No particular Course of Medical Study is at present marked out, but students are left to their own choice, or to the guidance of any Medical friend whom they may consult', with the consequence that an 'indiscriminate and erroneous mode of prosecuting Medical knowledge' was often followed by students.[73] While a larger proportion of medical students than Arts students chose to graduate, and it was common for over a hundred students to graduate every year in the early nineteenth century, this still only constituted a minority of those who came to Edinburgh to study medicine. The vast majority of those who chose to graduate were English, Irish or from the colonies and had come to Edinburgh for a year to finish their medical education after studying elsewhere.[74] Of the 160 students who graduated with an MD in 1827, fifty were Scottish, forty-six English and thirty-six Irish. Most of the others were from the West Indies, Canada and other colonies, although there were also a handful of students from outwith the British Empire.[75]

The report of the Universities Commission of 1826 contains some interesting evidence on the attitude of the Edinburgh medical professors towards the tradition of studying in the Arts faculty before going on to study medicine at the Scottish Universities. While this had been normal for Scottish students, there was no formal requirement to do so, and those from outside Scotland generally did not follow this path, but commenced their medical studies straight from school. The large number of such students studying medicine meant that the professors had a strong incentive to resist the imposition of any requirement to have completed an Arts degree, or any equivalent general education, before proceeding to medicine. In a paper submitted to the Commission by William Alison, professor of the theory of physic, and approved by all the medical professors except Andrew Duncan, professor of materia medica, it was made quite clear that such a general education in literature and science was not only unnecessary but positively harmful for medical students:

> Every Medical man has besides to acquire habits of business, observation of mankind, and a knowledge of the world. These acquirements of themselves make up to many Medical men for the want of Scientific knowledge, but the knowledge of all the Sciences cannot make up for the want of those; and, in general, I believe, we may say, that the habits of a Student who has gone through a very long and varied course of Literature and Science, are not those which will fit a man for that kind of intercourse with the world by which a Physician may live.[76]

Duncan, when asked by the commissioners 'What particular objection occurs to you, to requiring evidence of having attended the Philosophical Classes in the Universities, previously to taking the Degree of Medicine?', gives a very enlightening reply:

> The objection which occurs to me, and which is in my own mind quite conclusive against it, is, that a very large proportion of our Graduates come from England and Ireland, where, during the period of their life before they begin their professional study, they have no means of attending those branches in a University, although they may acquire moderate knowledge of them in schools, both public and private, and from their own private exertions. If we were to require evidence of their having attended them in Universities, in addition to the four years of professional study, the Degree here could not be obtained without a residence of not less than six or eight years – the effect of which would be, that these gentlemen would take no Degree at all.[77]

The issue at stake here seems to have centred on the hostility on the part of the Medical faculty to any proposal by the Universities Commission that any prior evidence of educational attainment be made compulsory for students of medicine. As professors relied on the fees they could charge for their courses for all or most of their income, anything that might shrink class sizes was to be forcefully resisted.

The commissioners were not swayed by these arguments, however. They dismissed the conclusions of Alison's submission, arguing instead that 'the preparatory education for which some contend does not interfere in the slightest with degree with the Medical – it only tends to make the practitioner a more enlightened man, and it is not easy to see how the acquisition of it should have the effect which Dr. Alison and the Faculty would assign to it'.[78] After consulting with a number of leading practicing physicians, including John Abercrombie (1780–1844), they decided on the contrary that '[t]he conclusion to be deduced seems unquestionably to be decided in favour of a superior preliminary education to that which is now required'.[79] This preliminary education could, in their opinion, be obtained either through the traditional route of completing a Master of Arts degree, or by other means 'without the slightest hardship'. The commissioners were in no doubt that the character of the medical profession would be improved if all practitioners received a thorough literary and scientific education in addition to their professional training.

## EDINBURGH'S EXTRA-MURAL ANATOMY SCHOOLS

The deficiencies in the teaching of anatomy by Alexander Monro, tertius helped to create a lively demand for extra-mural classes in the subject to supplement his uninspiring lectures.[80] Mudie remarks of the situation in the mid-1820s that:

> for the fragments of the medical school that remain, the Athens is almost wholly dependent upon private lectures; that the students pay their fees and enter

their names at the college, not with any view of attending the classes there, but because the fees and entries are necessary for the ceremony of graduation.[81]

The need for additional tuition must have been keenly felt, as students were obliged to pay to attend the classes of the university professors whether they considered them worth the price or not, and so would end up paying twice over for subjects for which they chose to attend extra-mural classes. This also worked to ensure that the classes at the extra-mural schools were highly responsive to student needs and expectations. Alexander Grant considered the competition provided by the extra-mural schools to be a positive boon for medical education in the city. He commented that the Medical faculty of the University was 'surrounded by extra-mural rivals, who have kept its Professors up to the mark, and sometimes eclipsed them'.[82] As an added attraction they often featured a certain degree of showmanship, as is evident in the successful career of the flamboyant extra-mural anatomy lecturer Robert Knox. In 1822 Bower listed twenty-three extra-mural teachers offering their services to Edinburgh's medical students.[83] The Universities Commission does not seem to have seen the reliance of students on the extra-mural schools as a problem. They suggested not only that the chair of anatomy should be disjoined from surgery – both were held by Monro at the time – but that it seemed 'doubtful whether [practical anatomy] could be taught within the College, or might not be left to private Lecturers, who enjoy advantages for teaching it which a Professor might not possess'.[84] The commissioners did not clarify what these advantages were, but it can be conjectured that more ready access to cadavers for dissection might have been one important consideration. I will concentrate my attention here principally on the school run by John Barclay. Not only was this the most successful of the extra-mural anatomy schools in Edinburgh, but two very significant figures, Robert Knox and Robert Grant, both taught there, and Knox went on to take over the running of the school after Barclay's death.

Barclay taught anatomy in Edinburgh from 1797 to 1825. As a young man he had studied for the ministry at the University of St Andrews before coming to Edinburgh to study medicine, and he remained a deeply religious man all his life. He was a great success as a lecturer; it has been estimated that Barclay's classes accommodated four or five hundred students a year.[85] Apart from the quality of his lectures, Barclay's popularity also owed something to the close proximity of his lecture hall to the Royal Medical Society and Surgeon's Hall, making it very convenient for students.[86] In order to cope with demand, he had to give two lectures a day, one in the morning and one in the evening. Such was Barclay's reputation as a comparative anatomist, that, as noted above, in 1817 there was a move to create a chair of comparative anatomy at the University for him, which was defeated by the professors of the university, jealous of their prerogatives.

When Barclay became ill shortly before his death in 1826 his classes were taken over by Robert Knox, who subsequently became the proprietor of the school. Knox was a colourful and controversial figure who graduated MD from the University of Edinburgh in 1814. During his studies he had attended Barclay's extra-mural anatomy lectures. He inherited radical political views from his father, who had been a member of the Jacobin-inspired Friends of the People. After his studies he spent some years as an army doctor. He served at the Cape of Good Hope between 1817 and 1820, an experience that helped him develop his unorthodox views on race and his strongly anti-colonial attitudes. In 1821/2 he spent a year studying in Paris, where he attended courses by Cuvier, Henri Marie Ducrotay de Blainville and Dominique Jean Larrey, and most importantly for the development of his opinions on comparative anatomy, Geoffroy Saint-Hilaire.[87] During his time in Paris he became a steadfast convert to Geoffroy's brand of philosophical anatomy. Back in Edinburgh, he was an active member of the Wernerian Natural History Society and published widely on various topics in comparative anatomy. Knox's biographer and former student, Henry Lonsdale, notes that during this period '[h]is studies at the Museum of Natural History made him known to Professor Jameson, who was glad to receive the aid of a promising naturalist for his *Quarterly Philosophical Journal* [sic]'.[88] Jameson regularly entrusted Knox with rare specimens of exotic animals for dissection, which formed the basis of a number of papers given by Knox to the Wernerian Society. Knox also had the opportunity to renew his acquaintance with Barclay, and on 2 March 1825 Knox signed an agreement of partnership in Barclay's extra-mural anatomy school.[89] The high quality, colourful style and often controversial content of Knox's classes made him a very popular lecturer with medical students. Knox was very much the showman. Even his manner of dress was calculated to create a strong impression on his students. In his biography Lonsdale recalled how 'Knox, in the highest style of fashion, with spotless linen, frill, and lace, and jewellery redolent of a duchess's boudoir, standing in a class amid osseous forms, *cadavera*, and decaying mortalities, was a sight to behold, and one assuredly never to be forgotten'.[90] Lonsdale's description emphasised the theatricality of Knox's performances, which so captivated students that they could hardly be awakened from their 'half-entranced condition when the "To-morrow, gentlemen", and the graceful bow, showed the falling of the curtain'.[91] Smallpox contracted during childhood had left Knox blind in one eye and with a severely pockmarked

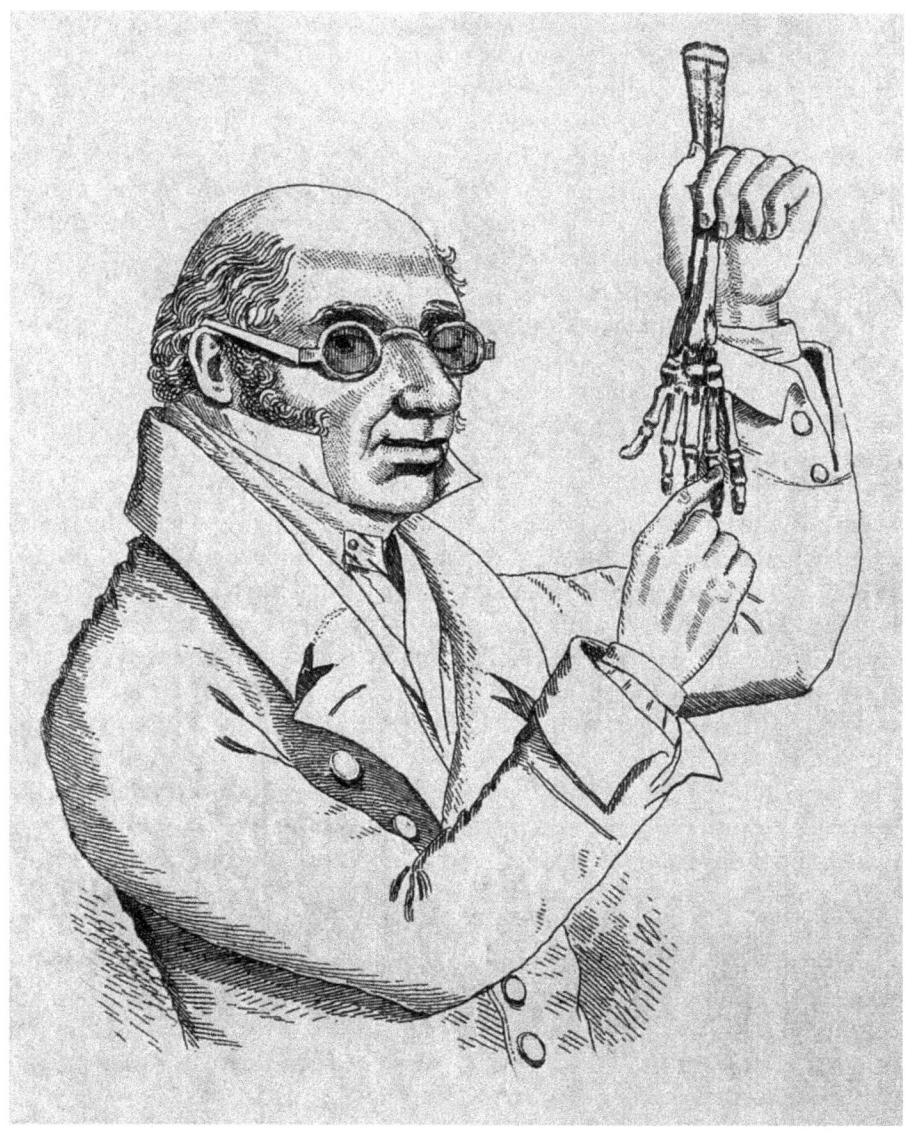

**Figure 2.2** Line engraving of Robert Knox. Credit: Wellcome Collection. CC BY.

face, and it is possible that his extravagant manner and showy appearance were his way of demonstrating his defiance of these physical shortcomings.

Whatever the reason, his striking and theatrical persona seems to have contributed greatly to his success as a lecturer. In only three years Knox increased the attendance of 300 students at Barclay's school to more than

500. This figure suggests that around two-thirds of Edinburgh medical students attended Knox's course, making it the largest anatomy class in Britain.[92] In 1825 he also became a fellow of the Royal College of Surgeons of Edinburgh, where he was appointed the conservator of their museum the same year. Barclay died in 1826, leaving Knox as sole proprietor of his anatomy school. Unfortunately for Knox, his involvement in the scandal surrounding Burke and Hare's infamous series of murders, committed to meet the need of the anatomy schools for fresh cadavers, cast a blight over Knox's once flourishing career. Knox had bought twelve of the bodies from the murderers, and his name became indelibly associated with their crimes in the public mind. His career subsequently went into a long decline that led him to move to London in the early 1840s, where he supported himself and his family largely by writing and giving public lecture tours until his death in 1862.

We have seen that the University of Edinburgh in the early nineteenth century had a dynamic intellectual culture, where students were challenged to debate and defend their own ideas rather than being passive recipients of knowledge handed down from on high. This was reflected in the manner of teaching of many of the professors as well as in the lively culture of student societies, where the latest scientific and philosophical ideas were fearlessly debated. The student coming to study at Edinburgh found himself in the midst of a student body that was socially and religiously diverse in a way that would have been unimaginable at Oxford or Cambridge. The professors too, despite the power of patronage networks and endemic nepotism, were a more diverse body than might have been expected. In addition, they had astonishing liberty regarding what and how they taught. This tolerant and liberal atmosphere owed something to the unique historical development of the Scottish universities and much to the influence of the Moderate literati who dominated university life in the last decades of the eighteenth century and whose influence lingered long into the nineteenth. This vibrant, tolerant environment was a world away from the paternalistic, clerical atmosphere at the contemporary English universities.

The liberty afforded to both students and professors alike bred a receptiveness to new ideas. As James Lorimer remarked, 'If there is one peculiarity in the intellectual character of our countrymen, that we specially prize, it is that openness and freshness of mind which is ready to receive new truth, wheresoever it may come'.[93] This openness and freshness of mind was prized not only by the local students, but also exerted a powerful influence on those students that came to study in the city from other parts of the British Isles and beyond. As we will see in subsequent chapters, new ideas about the natural world flooded into Edinburgh via the scientific journals published in the city, which acted as conduits for the latest ideas from France and Germany. They were also spread by the teachers at some of the extra-mural schools, such as Robert Knox and Robert Grant, both of whom had studied in Paris, at that time a veritable crucible of revolutionary new ways of thinking about the natural world as much as it was the epicentre of European political developments.

In his dedication to teaching, his willingness to discuss and debate with his students and his openness to new ideas, Robert Jameson was a model Edinburgh professor, setting a standard that few of his fellows could rival. Charles Darwin's dislike of Jameson for too long blotted his reputation. More recent scholarship has, however, begun to reveal a very different and more complex figure. In the next chapter we will be taking a closer look at Jameson, at his natural history class, and at Edinburgh's lively natural history circles in general.

## NOTES

1. [Mudie], *The Modern Athens*, p.221. In the final decades of the eighteenth century James Gregory was a professor of medicine at Edinburgh, Alexander Monro, secundus was professor of anatomy and Joseph Black was professor of chemistry.
2. For a discussion of the religious qualifications at the English universities in this period, see Cantor, *Quakers, Jews and Science*, pp. 822–94.
3. Smollett, *Expedition of Humphrey Clinker*, p.232 (original emphasis, here and elsewhere).
4. Sher, *Church and University in the Scottish Enlightenment*.
5. Smollett, *Expedition of Humphrey Clinker*, p.232.
6. Emerson, 'The contexts of the Scottish Enlightenment', p.16.
7. Sher, *Church and University in the Scottish Enlightenment*, p.151.
8. Ibid., p.114.
9. Morrell, 'The University of Edinburgh in the late eighteenth century', p.158.
10. Stewart, 'Life and writings of William Robertson', p.lxxxiii.
11. Morrell, 'Science and Scottish University reform', p.39.
12. Ibid., p.42.
13. Grant, *Story of the University of Edinburgh*, vol.1, p.265.
14. Ibid., p.277.
15. Anderson, *Universities and Elites*, p.5.
16. Horn, *Short History of the University of Edinburgh*, p.65.
17. Scottish Universities Commission (1826), *General Report of the Commissioners*, p.73.
18. Bower, *Edinburgh Student's Guide*, p.xxiii.
19. Anderson, *Universities and Elites*, p.24.

20. Scottish Universities Commission (1826), *General Report of the Commissioners*, pp.73–4.
21. [Lockhart], *Peter's Letters to his Kinsfolk*, vol.1, pp.191–2.
22. Horn, *Short History of the University of Edinburgh*, pp.119–20.
23. Playfair, inaugural address (1858), quoted in Davie, *Democratic Intellect*, p.21.
24. Scottish Universities Commission (1826), *Report Relative to the University of Edinburgh*, p.47.
25. Jardine, *Outlines of Philosophical Education*, p.vii.
26. Scottish Universities Commission (1826), *Report Relative to the University of Edinburgh*, p.97.
27. Scottish Universities Commission (1826), *Abstract of the General Report*, pp.8–9.
28. Ibid., p.9.
29. Cockburn to Kennedy, 27 February 1826, in Cockburn and Kennedy, *Letters Chiefly Concerned with the Affairs of Scotland from Henry Cockburn to Thomas Francis Kennedy*, p.138. For an excellent account of Hope's teaching see Morrell, 'Practical chemistry at the University of Edinburgh'.
30. Scottish Universities Commission (1826), *General Report of the Commissioners*, p.36.
31. Morrell, 'Science and Scottish university reform', p.52.
32. [Brewster], Review of Babbage's *Reflections on the Decline of Science in England*, p.326.
33. Grant, *Story of the University of Edinburgh*, vol.1, p.302.
34. Anderson, 'Scottish university professors', p.39.
35. Scottish Universities Commission (1826), *Report Relative to the University of Edinburgh*, p.36.
36. Anderson, 'Scottish university professors', p.50.
37. Morrell, 'The Leslie affair'.
38. Gordon, *Home Life of Sir David Brewster*, p.61.
39. Leslie, *Experimental Inquiry into the Nature and Propagation of Heat*, p.521.
40. Morrell, 'Science and Scottish university reform', p.42.
41. Cockburn, *Memorials of His Time*, pp.95–6.
42. [Mudie], *The Modern Athens*, p.220.
43. Jacyna, *Philosophic Whigs*, p.52.
44. Cockburn, *Memorials of His Time*, p.80.
45. Morrell, 'Science and Scottish university reform', pp.43–4.
46. Scottish Universities Commission, *Abstract of the General Report*, p.36.
47. Anderson, 'Scottish university professors', p.35.
48. Scottish Universities Commission, *Report Relative to the University of Edinburgh*, p.74.
49. Brougham, *Life and Times of Henry Lord Brougham*, vol.1, pp.81–2.
50. Jacyna, *Philosophic Whigs*, p.60.
51. Mackintosh, *Life of the Right Honourable Sir James Macintosh*, p.29.

52. Anderson, *Universities and Elites*, p.6.
53. Bower, *Edinburgh Student's Guide*, p.6.
54. Morrell, 'The University of Edinburgh in the late eighteenth century', p.169.
55. Scottish Universities Commission, *Report Relative to the University of Edinburgh*, p.88.
56. Horn, *Short History of the University of Edinburgh*, p.76.
57. The issue of tests at the Scottish universities was later to become a hotly contested one in the changed climate of the 1840s.
58. Davie, *The Democratic Intellect*, p.3.
59. Ibid., p.5.
60. Ibid., p.26.
61. Scottish Universities Commission, *Report Relative to the University of Edinburgh*, p.95.
62. Ibid., p.96.
63. Brougham, *Life and Times of Henry Lord Brougham*, vol.3, p.4.
64. Horn, *A Short History of the University of Edinburgh*, p.108.
65. Chitnis, 'Medical education in Edinburgh', p.174.
66. Charles Darwin to Caroline Darwin, 6 January 1826, in Darwin, *Correspondence of Charles Darwin*, vol.1, p.25.
67. Darwin, '1876 May 31 – Recollections of the development of my mind and character', in Darwin, *Autobiographies*, p.22.
68. Browne, *Charles Darwin: Voyaging*, p.56.
69. Chitnis, 'Medical education in Edinburgh', pp.174–5.
70. Ibid., p.176.
71. Ibid., p.177.
72. Scottish Universities Commission, *Report Relative to the University of Edinburgh*, p.79.
73. Ibid., p.105.
74. Horn, *Short History of the University of Edinburgh*, p.46.
75. Grant, *Story of the University of Edinburgh*, vol.1, p.329.
76. Scottish Universities Commission, *Report Relative to the University of Edinburgh*, p.102.
77. Ibid., pp.103–4.
78. Ibid., p.108.
79. Ibid., p.105.
80. Kaufman, 'John Barclay', p.95.
81. [Mudie], *The Modern Athens*, pp.221–2.
82. Grant, *Story of the University of Edinburgh*, vol.1, p.293.
83. Bower, *Edinburgh Student's Guide*, pp.142–6.
84. Scottish Universities Commission, *Report Relative to the University of Edinburgh*, p.105.
85. Browne, *Charles Darwin: Voyaging*, p.57.
86. Rosner, 'Barclay, John'.
87. Taylor, 'Knox, Robert'.

88. Lonsdale, *Life and Writings of Robert Knox*, p.36. Presumably Lonsdale is referring here to the *Edinburgh Philosophical Journal*.
89. Lonsdale, *Life and Writings of Robert Knox*, p.44.
90. Ibid., p.126.
91. Ibid., p.148. The work of Iwan Rhys Morus fruitfully explores the role of performance in nineteenth-century science. See in particular Morus, 'Worlds of wonder' and 'Placing performance'.
92. Rehbock, *Philosophical Naturalists*, p.44.
93. Lorimer, *Universities of Scotland Past, Present and Possible*, p.38.

# 3

# *Natural History in Edinburgh, 1779–1832*

In the early nineteenth century Edinburgh had the most prestigious chair of natural history in Britain, largely thanks to the efforts of its remarkable third professor, Robert Jameson. Natural history circles in the city in this period maintained a lively interchange of ideas with other major European centres of innovation in the natural sciences. Many of the city's natural historians had studied abroad and had personal connections with some of the most important figures in European science. It should therefore be no surprise that it was also the city in Britain where the latest radical ideas on the natural world were most readily embraced, or that some of its natural historians went on to develop these ideas in new directions. In this chapter I will be exploring in more detail the institutional context that made this remarkable efflorescence possible. This will necessitate exploring those important contexts in which new evolutionary ideas were received, debated and developed, from the chair of natural history and the natural history circles surrounding its professor, to the ferment of activity in its student societies and the journals published in the city that broadcast these new ideas to the wider scientific community. First I will explore natural history in Edinburgh in the final decades of the eighteenth century, in the years before Robert Jameson ascended to the chair of natural history in 1804, focusing in particular on John Walker (1731–1803), Jameson's predecessor in the chair, and William Smellie (1740–95), Walker's great rival. Then I will look at the university, its chair of natural history and the city's extra-mural anatomy schools as they were in the early decades of the nineteenth century, before turning to its natural history societies and the journals published there.

## NATURAL HISTORY IN EDINBURGH IN THE LATE EIGHTEENTH CENTURY

Before exploring the Edinburgh of the early decades of the nineteenth century, it is worth going back to the last decades of the previous century to examine the state of natural history in the city in those years. The chair of natural history at Edinburgh was occupied by John Walker from 1779 until his death in 1803. Walker himself graduated from the University of Edinburgh in 1749 and had then became a minister in the Church of Scotland. From his parishes first in Glencorse and then from 1764 in Moffat he pursued his interest in natural history, which brought him into contact with two men who were to become important patrons, William Cullen, professor of chemistry at Edinburgh, and Henry Home, Lord Kames. It was with their help, and that of other influential acquaintances, that he was able to secure the post of regius professor at the University of Edinburgh in 1779. His main rival for the chair when it became vacant on the death of Robert Ramsay (1735–78), Edinburgh's first professor of natural history, was William Smellie. Smellie was a successful Edinburgh publisher and the 'editor and author of works that helped to define the Scottish Enlightenment', as well as a founding member of the Royal Society of Edinburgh.[1] He too owed his success in part to the patronage of Kames, who encouraged him to write his two-volume *Philosophy of Natural History* (1790–91). Smellie also produced a translation of the *Natural History* (1780) of George-Louis Leclerc, Comte de Buffon (1707–88), which was still being reprinted well into the nineteenth century.

Walker's chief interest was mineralogy. He had little time for the speculative theories of the earth that formed an important component of natural history discourse in the period. As Eddy has noted in a recent study, 'Buffon's cosmology typified two things that most irritated him; first, unconfirmed and therefore potentially erroneous information, and second, a love of theoretical systems'.[2] He fully accepted a relatively recent date for the creation of the earth, in line with orthodox interpretations of the Biblical chronology, and his approach to his subject was largely non-historical. Despite his hostility towards system building, on the few occasions he did deign to speculate on the history of the globe, he seems to have believed that the oldest, primary strata of the earth had precipitated out from an aqueous solution, which, however, he did not equate with the Biblical flood.[3] He shared this belief with Abraham Gottlob Werner (1749–1817), with whom Robert Jameson would later study in Freiberg. He may well have come to this opinion independently, however, as Eddy has found no reference to Werner in any of Walker's lectures or personal notes.[4]

Walker appears to have been a dedicated and conscientious professor who took his teaching responsibilities extremely seriously. In addition to his formal lectures, he organised field trips for his students and held tutorials in the Natural History Museum of the university, of which he was also the keeper. He also encouraged his students to found societies, including the Natural History Society in 1782 and the Chemical Society in 1785. The interest he showed in stimulating the enthusiasm of his students through these additional activities is likely to have contributed to the popularity of his class with students from a wide variety of backgrounds. The largest single group attending his lectures were medical students, making up around half of the total number. The rest were students from the other faculties of law, divinity and arts, aristocrats with an interest in natural history or individuals with a professional interest in the subject matter of his course, such as apothecaries or jewellers.[5] The matriculation policy of the university, which allowed students to pay for their studies on a class-by-class basis and to attend lectures that were not required for their degrees, may help to explain the diversity of the audiences his lectures attracted.

A concept that dominated natural history in Edinburgh as it did almost all discourse on the natural order in the eighteenth century was the idea of the 'great chain of being'.[6] This principle was based on the idea that it was impossible that a deity whose attributes were infinite would not create a universe where all potentiality for being was realised. In the classic formulation of Walker's Swiss contemporary Charles Bonnet (1720–93), of whose work Walker was certainly aware, '[b]etween the lowest degree and the highest degree of physical or spiritual perfection there is an almost infinite number of intermediate degrees. The series of these degrees constitutes the *Universal Chain*.'[7] Thus nature was envisaged as an infinite hierarchy of beings ascending from inanimate matter to the deity himself, as is so memorably illustrated in Pope's *Essay on Man*, in which each link in the chain is described as locked in place in such a way that: 'Where, one step broken, the great scale's destroyed'.[8] This, 'an absolutely rigid static scheme of things', was fundamentally incompatible with any suggestion of change or progress in the natural world.[9]

One prominent Edinburgh natural historian who wrote at length on the Great Chain of Being was William Smellie. In the final chapter of volume I of his *Philosophy of Natural History* (1790), which is entitled 'Of the progressive scale of animals', he wrote that '[t]here is a graduated scale or chain of existence, not a link of which, however seemingly insignificant, could be broken without affecting the whole'.[10] The natural order, at least in the case of living things, did not allow for any gaps, and 'the gradations from one species to another are so imperceptible that to discover the marks of their discrimination requires the most minute attention'.[11] Organisms were locked into an ascending natural order that was complete and unchanging.

Although there was little love lost between Smellie and Walker as a result of their rivalry for the chair of natural history,[12] they were of one mind on the question of the order of nature. In a set of student's lecture notes from Walker's course in 1791 we have the following exposition of the chain of being: 'There is undoubtedly a continued chain in nature from its lowest subject up to the human species, & which it is to be supposed proceeds from him to his maker, all being linked as in the moral world by the most beautiful & regular gradation; for nothing is more certain than the maxim "Natura nunquam fit saltus."'[13]

The continuity that Smellie and Walker imagined as an attribute of the chain of being has another interesting consequence for Smellie in his *Philosophy of Natural History*. The seamless unity of the natural order can also be used to explain the obvious family resemblances between some species. Smellie noted, for example, that 'in the creation of animals, the Supreme Being seems to have employed only one great idea, and, at the same time, to have diversified it in every possible manner, that men might have the opportunity of admiring equally the magnificence of the execution and the simplicity of the design'.[14] It is likely that Smellie was familiar with the ideas on unity of plan of Buffon, whose work he had translated; in volume 8 his translation of Buffon's *Histoire Naturelle* we find the following passage: 'in all of them [vertebrate animals] he found a solid structure composed of the same pieces, and nearly situated in the same manner. This plan proceeds uniformly from man to ape, from ape to quadrupeds, from quadrupeds to cetaceous animals to birds, to fishes, and to reptiles.'[15] In Smellie's vision of the natural world, this unity of plan was a natural consequence of the imperceptibly close gradations that existed between the links in the chain of being. Following this line of reasoning, he went on to claim in his *Philosophy of Natural History* that 'Man, in his lowest condition, is evidently linked, both in the form of his body and the capacity of his mind, to the large and small orang-outangs'.[16] However, this is emphatically not an evolutionary connection, but simply the result of close proximity in the progressive scale of being. To suggest that the orang-utan could one day become human would be to imply that it was possible to break the chain of being, destroying the integrity of the natural order.

In his lectures, Walker made very clear that he considered species to be immutable. In a student's notes from his lectures in 1782 we find him stating that it is 'probable that no species of P[lant] or An[imal] changes into another & no species lost or new formed'.[17] In 1790 he stated categorically that '[b]oth in the Vegetable and Animal Kingdom there are a certain number of distinct species which have remained without addition or diminution since the creation'.[18] In 1797 he is still telling very much the same story, although in even more emphatic terms: 'There therefore appears to be no Species of Plants or Animals entirely lost, or any new Species formed.

The Transmutation of Species either in Plants or Animals, is a Vulgar Error.'[19] That Walker found it necessary to refute transformism, and to continue to do so from the early 1780s through to the later 1790s, shows that he was well aware of the idea, and felt it was well-known enough to require categorical refutation.

For Walker, therefore, species were fixed for all eternity. Nonetheless, as we will see in the next chapter, the same was not necessarily true for varieties of plants and animals, which he defined as 'beings belonging to a species, and differing from it in some trifling circumstance'.[20] Such varieties were simply the result of conditions of life, such as the climate or the availability of food, acting on individuals of the species. Walker's ideas about the nature of varieties are likely also to have been shaped by his reading by Buffon, whose works Walker cited regularly in his lectures. In volume 6 of his *Histoire Naturelle*, quoted here in Smellie's translation of 1785, Buffon gave an account of the effect of the conditions of life on varieties of animals.

> And, if we examine each species in different climates, we shall find sensible varieties both in size and figure. These changes are produced in a slow and imperceptible manner. Time is the great workman of Nature. He moves with regular and uniform steps. He performs no operation suddenly; but by degrees, or successive impressions, nothing can resist his power; and those changes which at first are imperceptible, become gradually sensible, and at last are marked by results too conspicuous to be misapprehended.[21]

It is not clear whether Buffon is here proposing a mechanism for the development of new species as well as varieties, although it is quite evident that Walker was thinking only of the generation of varieties. In any case, the manner and mechanism of the transmutation would appear to be the same.

In contrast to Walker, Alexander Monro, secundus, the professor of medicine and anatomy at the University of Edinburgh, was thoroughly hostile to Buffon's ideas on even the transmutation of varieties. In his anatomy lectures for 1774/5 he gave a summary of Buffon's account of the development of new varieties of dog through the influence of their conditions of life. He went on to challenge his ideas, claiming that '*Buffon* refutes himself by the very accurate enumeration he gives. If the variety of Dogs depends upon the Circumstances he supposes, how comes he to find there is a certain No. only, the No. should have been endless, considering of the succession of Ages.'[22] Again, a decade later, in a set of notes from 1786, Monro still began his lectures on comparative anatomy with a refutation of Buffon's views on the mutability of varieties. Once again drawing on the example of breeds of dogs, he claimed that 'authors in general as

Buffon go far wrong in deeming varieties much less constant than they really are for they attribute the many varieties of dogs to the difference of climate & other external causes'. He also denied that the varieties could interbreed successfully in the long term: 'I well know that mongrels are prod.$^d$ in dogs but these I think die out in a very few Generations.'[23] Even if not everyone accepted Buffon's theories on the production of new varieties through the influence of their conditions of life, it is clear that they were widely known, and that those who disagreed with them considered them at least worthy of refutation.

In Carl Linnaeus' *Disquisitio de sexu plantarum* of 1760 he suggested not only that hybridisation was possible, but that inter-specific hybrids could be fertile. He went on to propose that new species could arise in this manner. In his *Disquisitio* he wrote: 'It is impossible to doubt that there are new species produced by hybrid generation.'[24] Walker seems to have been aware of these views, as in a lecture of 1790 he stated that 'Linnæus, lately thought, that all plants changed to their present dissimilar forms during the process of time from a single species. He and Bonnet have thought too that plants may and do continue to start up during the course of Ages. An Opinion in which few will be willing to follow him.'[25] Presumably it was Linnaeus' theory of the production of new species through hybridisation that Walker had in mind here, although there is the suggestion that Linnaeus went further than this to propose a common origin for all species of plants. In 1797 Walker explicitly denied that inter-specific hybrids could ever be fertile, stating that 'if Mules were fertile the whole Vegetable and Animal Creation, would run into Confusion and Disorder, so that this Infertility of the Mules, may be presumed to be intended by the wise Creator of the Universe, to preserve the Order and Regularity, which is every where so conspicuous over the Globe'.[26] The infertility of hybrids is therefore presented as a mechanism of divine providence for keeping the chain of being in order. Monro also emphasised the reduced fertility of hybrids even among different breeds of dog, claiming that any hybrid breed would 'generate a certain No. of Times, but after that the spurious breed wears out, & hence the varieties come to be marked so plainly'.[27] It seems that on this matter Walker and Monro spoke with one voice.

At times Walker even seemed to doubt whether inter-specific hybridisation was possible at all. We know from the 1790 lecture notes that Walker knew about Linnaeus' claim, published in 1749, that the newly discovered species *Peloria* was a hybrid, but that he insisted that this was not the case and that it was only a remarkable variety of an existing species.[28] Walker went on to recount how Daniel Solander, the Swedish naturalist and disciple of Linnaeus, had personally shown him a supposed example of a hybrid between *Mirabilis talapa* and *Mirabilis longiflora* plants that had been sown in the same bed, but dismissed his claims as he regarded the parent plants as varieties of the same species. He was also aware of the experiments on hybridisation in plants performed by Joseph Gottlieb Koelreuter, who in 1761 produced the first well-authenticated inter-specific hybrid.[29] Again he dismissed these, as in his opinion the parent plants were 'only varieties tho delivered by Linnæus as different species'.[30] Although seemingly determined to reject all claims to have produced a race of hybrid plants, he was, however, prepared to concede that 'it would appear that the production of hybrid plants requires more experiments before any conclusion can be formed'.[31]

While Walker doubted the possibility of inter-specific hybrids, for him hybridisation between different varieties of the same species could take place and produce fertile progeny. At the end of lecture 42 of his course in 1790, Walker has something very intriguing to say about hybrids between different varieties of peacock:

> It is certain likewise that in some animals a variety will miss one generation and take place in the next. I have seen a white peacock and pea hen, which, when they bred produced peacocks and pea hens of the common sort, yet these, when they bred produced white peacocks and pea hens, but here Gentlemen I find the hour is elapsed.[32]

Walker does not offer any explanation for this phenomenon, which is presented merely as an anecdote to round off his lecture. Pierre Louis Maupertuis (1698–1759) had addressed the question of heredity in cases such as this in his *Vénus Physique* (1745), where he discussed the inheritance of albinism in humans at length.[33] We know that Walker had read some of the works of Maupertuis, or had at least heard about them at second hand; in one of Walker's lectures in 1797 he made a reference to Maupertuis' speculations about the possible impact of comets on the history of life on earth in his 'Essai de la Cosmologie' (1751).[34] However, if Walker knew of Maupertuis' ideas on heredity, or their relevance to the phenomenon he describes, we have no evidence that he said anything about it in his lectures.

In his course Walker gave a detailed account of different theories of generation. First he dealt with spontaneous generation, or as it was generally known at the time, 'equivocal generation', a concept that he traced back to Aristotle. Equivocal generation was the idea that life could arise spontaneously from non-living matter as well as from a previous generation of living beings; to quote some of the examples given by Walker, 'silk worms were generated from putrid mulberry leaves', and 'most other insects, from corrupted vegetables and animals'.[35] In the notes from his lectures in 1782 he said that the 'Doctrine of equivocal Gen[eration is] to be reviewed and rejected & that of univocal established'.[36] In 1791 he continued to declare in more or less the same words that 'The doctrine of Equivocal generation is here to be considered & rejected upon review'.[37] He went on to discuss and reject the opinions of Buffon on generation, which he presented in the following fashion:

> 1st That every where in nature there are certain Particulars [sic] Organiques.
>
> 2d That there are in nature certain external and internal moults [sic] in which these particles are formed into living existences and
>
> 3d That there is in nature a force productive sufficient to bring to life the particles fashioned by the Moults, thus, this differs very little from the Hypothesis of Equivocal generation[.][38]

After this description of Buffon's system, he went on to reject his ubiquitous organic molecules as chimeras. He finally settled on a theory he attributed to William Harvey (1578–1657), now best known for his work on the circulation of the blood, as closest to the truth, stating that the most likely explanation of generation was 'that a mixture of male and female liquors formed the Fœtus'.[39] He quoted with approval Harvey's dictum 'omne animal ex ovo', thereby denying the possibility of equivocal generation.[40] However, he also claimed in his 1790 lectures that 'Bonnet and Haller have proved that the embryo of the animals is in the female'.[41] The theory of Bonnet and Haller referred to here was a version of preformationism, in that it suggested that the adult form of the animal was contained within the egg of the mother, complete in all its parts, although tiny. There is clearly some contradiction here between the epigenetic argument for the formation of the embryo from a mixture of the 'liquors' of the two parents and the idea that the embryo derived entirely from the female. Which of these two mutually incompatible alternatives reflected Walker's own views is not clear. However, his ideas did not appear to have changed much by 1797, when he was still declaring himself in agreement with the opinions of Harvey and hostile to the idea of equivocal generation, which was central to the theories of many transformist thinkers, such as his contemporary Erasmus Darwin.[42]

## ROBERT JAMESON AND THE CHAIR OF NATURAL HISTORY

Robert Jameson is the most important single figure in our story. He occupied several overlapping and interconnecting roles. As professor of natural history, keeper of the University's natural history museum, editor of the

*Edinburgh Philosophical Journal* and its successor the *Edinburgh New Philosophical Journal*, and the founder and perpetual president of the Wernerian Natural History Society he was the pivotal figure in Scottish natural history in the first half of the nineteenth century and one of the most influential figures in his field in Europe. He was at the centre of an extensive network of patronage that included many of the most

**Figure 3.1** Robert Jameson. Stipple engraving by J. Jenkins, 1847, after K. Macleay. Credit: Wellcome Collection. CC BY.

important early nineteenth-century natural historians, including figures such as Robert Grant, Robert Knox, William MacGillivray, John Fleming and David Brewster. He acted as an important patron in particular to his younger associates Grant and Knox, whose careers he played a large part in fostering. He also provided these two key figures in Edinburgh in the 1820s with a platform for their ideas through his editorship of the *Edinburgh Philosophical Journal*, the *Edinburgh New Philosophical Journal* and the *Memoirs of the Wernerian Natural History Society*. He was the author of several important geology texts, including *Elements of Geognosy* (1809). His connections in the natural history world and wide network of former students put him at the centre of an extensive web of correspondents across which information and specimens were exchanged around the globe. With around two hundred students in his natural history class at its peak, he was in a position to influence a very large number of individuals, many of whom went on to become significant figures in the scientific world.

Jameson, who had been one of John Walker's students, succeeded him as regius professor of natural history in 1804. He was to hold the chair for half a century, until his own death in 1854. He was also the keeper of the University's natural history museum, which he built from modest beginnings into one of the best and most extensive natural history collections in the country over the course of his career. However, his importance goes far beyond his chair at the University.

Like Walker, Jameson was not from an especially privileged background, being the son of a soap manufacturer from Leith. He had been an unenthusiastic student at grammar school, and had to be dissuaded by friends from going to sea rather than continuing his studies after he left school.[43] Instead, he became the assistant to the surgeon John Cheyne in Leith. According to the *Biographical Memoir* of Jameson written by his nephew Laurence Jameson, at this time Jameson became acquainted with Charles Anderson, who had produced a translation of Abraham Gottlob Werner's work on the origin of mineral veins, *Theory of the Formation of Veins*.[44] He also attended courses of lectures by Walker in 1792 and 1793. Jameson often accompanied Walker on dredging expeditions in the Firth of Forth. In 1793 he visited London, where he spent much time in the museums and met a number of leading naturalists, including Joseph Banks and other members of the Linnean Society.[45] This inspired him to give up his medical training to devote himself instead to natural history.

In 1793 Jameson also made a trip to Ireland where his interest in Werner's theories and rejection of the rival model of the history of the earth proposed by James Hutton (1726–97) were encouraged by the Irish

**Figure 3.2** Robert Jameson through the eyes of one of his students. A caricature of Jameson from an anonymous set of student notes dated 1831/2. Jameson, Notes on natural history lectures (1831/2), f.3 verso. Reproduced courtesy of National Museums Scotland.

geologist Richard Kirwan, who pointed out to him 'several strong fails [sic] against the Huttonian theory'.[46] By 1796 Jameson had fully embraced the Neptunian theory of the earth of Werner, which had much in common with the geological opinions that he would have heard from his friend and patron Walker. Evidence for this is to be found in two papers he read to the Royal Medical Society in that year, in which he expressed an uncompromisingly Wernerian view of the history of the earth.[47] He was to become the principal champion in Scotland of Werner's theories in the conflict that took place in the early decades of the nineteenth century between the disciples of Werner and the followers of Hutton. In 1800 Jameson travelled to Freiberg to study mineralogy and geology with Werner himself. On his return to Scotland Jameson assisted Walker, who was now old and in poor health, with his classes. While still a student he was given charge of the University's Natural History Museum by Walker.[48] When the old professor died in December 1803, Jameson succeeded him. As well as the leading Wernerian geologist in Britain, Jameson was also the most important interpreter of the ideas of the leading French geologist and comparative anatomist Georges Cuvier for a British audience. He contributed prefaces and notes to successive editions of Robert Kerr's translation of Cuvier's *Theory of the Earth* (1813). After Kerr's death in 1813 subsequent editions were entirely Jameson's work and became the most influential popular geology book in Britain between 1813 and 1830.[49]

Jameson's reputation has in the past been blighted by a number of remarks made by Charles Darwin regarding him. In a letter to J. D. Hooker in 1854 Darwin described Jameson as 'that old, brown dry stick Jameson'.[50] Hardly more flattering were Darwin's comments in his 'Recollections' in 1876, where he described Jameson's lectures as 'incredibly dull' and noted that 'the sole effect they produced on me was the determination never as long as I lived to read a book on Geology or in any way to study the science'.[51] Not everyone, however, agreed with Darwin. Indeed, it was the effusive praise of Jameson from Edward Forbes, Jameson's successor as professor of natural history at Edinburgh, that prompted Darwin's remark in his letter to Hooker quoted above. In his inaugural address, published in *The Scotsman*, Forbes said:

> Who, that in time past was his pupil and found pleasure in the study of any department of Natural History, can ever forget his enthusiastic zeal, his wonderful acquaintance with scientific literature, his affection for all his friends and pupils who manifested a sincere interest in his favourite studies. When, in after life, their fates scattered them far and wide over the world, some settling amid the civilised security of rural seclusion; some rambling to the far ends of the earth to sift and explore wild and savage regions; some plunging into the boiling and noisy whirlpool of metropolitan activity, none who remained constant to the beautiful studies of his pupilhood was ever forgotten by the kind and wise philosopher, whose quick and cheering perception of early merit had perpetuated tastes that might have speedily perished if unobserved and unencouraged.[52]

Although, it is, of course, unlikely that Forbes would have used his inaugural address to damn the memory of his former teacher and predecessor, there is no reason to believe that the sentiments expressed were not sincere. It would certainly have been difficult for Forbes to have provided a more glowing eulogy for Jameson than he did. Likewise, in the 'Biographical sketch of Robert Edmond Grant' that appeared in *The Lancet* in 1850, probably by Thomas Wakley, there is a reference to 'the highly attractive and invaluable lectures on Natural History of Professor Jameson'.[53] Although not everyone found Jameson an inspiring lecturer, despite his evident enthusiasm for his subject, he certainly seems to have been a popular one. Robert Christison later wrote of his time as a student of Jameson in 1816 that his 'lectures were numerously attended in spite of a dry manner, and although attendance in Natural History was not enforced for any University honour or for any profession, the popularity of his subject, his earnestness as a lecturer, his enthusiasm as an investigator, and the great museum he had collected for illustrating his teaching, were together the causes of his success'.[54]

Aside from the testimony of friends and former students, the undisputable popularity of Jameson's course surely provides ample testimony to his merits. The Scottish Universities Commissioners who interviewed Jameson in 1826 were more likely to be given an impartial verdict than figures such as Forbes, especially as they had sternly criticised Jameson for his management of the University's museum. However, they had nothing but praise for the excellence of Jameson's teaching and testified to the popularity of his course with students, reporting that: 'The average number of Students is stated at 200, being a great increase under the present very able and enlightened Professor – his first course of Lectures in the College having been attended only by 35.'[55] Bearing in mind that the natural history course was not a requirement for graduation for any degree from the University, and therefore was entirely optional, this was quite an achievement.

Not only does the report of the 1826 Scottish University Commission provide evidence for Jameson's qualities as a professor, it also provides a valuable insight into the practicalities of how he actually went about teaching his course in the mid-1820s. He appears to have inherited his style of teaching from Walker along with the outlines of his syllabus. According to the report, his class met for 'one hour each day, on five days of the week, for five months in the course of the Winter, and for one hour a-day during three months in the Summer Session'.[56] It went on to note that the 'mode of teaching in the Natural History Class is by Lectures and Demonstrations of the objects of Natural History; and with the view of impressing the details upon the minds of the pupils, the Professor makes it a practice to converse with them an hour before the Lecture, and very frequently after the Lecture'.[57] In addition to the lectures, Jameson also met with his students in the Museum three or occasionally six times a week for more informal discussions of subjects of natural-historical interest. He also took his students on regular field excursions. As Jameson, like the other Edinburgh professors, received most of his income from the fees from students who attended his course he had every reason to make it as attractive as possible, and added inducements such as free access to the museum and field trips must have helped him to do this. Jameson was clearly aware of the importance of such extras as a significant added attraction to potential students, and included a memorandum to the end of his syllabus for 1826 stating that:

> The Professor, still further to secure the attention of the Students, and also to afford them the means of becoming practically acquainted with many of the subjects explained during the course of the Lectures, makes along with them frequent excursions into the neighbouring districts; and when circumstances allow for it, even to distant parts of the country. These walking and sailing excursions form a peculiar and most useful part of the system followed in the Natural History Class. Although the Professor is generally accompanied in his expeditions by the greater number of his Pupils, no confusion arises; on the contrary, the information is equally shared by all, and a universal feeling of satisfaction and delight is the constant result of these peripatetic excursions.[58]

The museum played a central role in Jameson's teaching as well as greatly adding to his prestige. A large proportion of Jameson's surviving correspondence relates to the acquisition of objects for his collection, many of them procured for him by former students and other contacts around the world. Graduates of the University of Edinburgh's medical school took up posts in all corners of the British Empire and beyond, and Jameson was keen to ensure that any objects of natural-historical interest they came by would find their way to his museum. With this end clearly in view, he dedicated part of the zoology section of his lecture course to encouraging his students to collect specimens and send them to him for the museum. In his natural history syllabus for 1826, we find the following three items:

1. Instructions and Demonstrations as to the mode of collecting, preserving, transporting, and arranging objects of Natural History.
2. Collecting of Objects of Natural History strongly recommended.
3. Advantages of Travelling.[59]

Not content to offer much practical advice on collecting, preserving and shipping specimens to ensure their safe arrival, he also seems to have actively urged students to travel and collect as important and worthwhile activities in their own right. This advice was reflected in the subsequent lives and career choices of many of his students.

Jameson also granted access to the museum as a form of patronage, making specimens freely available to his friends and favourite students while denying access to those of whom he did not aprove. Robert Knox,

*Natural History in Edinburgh* 51

**Figure 3.3** An undated print showing Edinburgh College Museum as it would have appeared in Robert Jameson's time. A copy of the print was found among Jameson's papers (Edinburgh University Library, Gen. 1999/1/9). The creature under the display case on the left is the lynx reputed to roam the museum after dark. Reproduced courtesy of the Centre for Research Collections, Edinburgh University Library. CC BY.

for example, greatly benefited from access to the riches to be found in the University's natural history museum. Jameson regularly entrusted Knox with rare specimens of exotic animals for dissection, which formed the basis of a number of papers given by Knox to the Wernerian Natural History Society. According to the *Memoirs* of the Society for 1824, in that year alone Knox had reason to thank Jameson for access to specimens of an assortment of reptiles, an *Ornithorynchus* (duck-billed platypus), a chameleon and a collection of human skulls.[60]

While the Scottish Universities Commission had nothing but praise for Jameson's teaching, his management of the museum did come in for some criticism. His decisions to grant or deny access to the collections could be arbitrary and based on personal prejudices. The Commission had received several complaints about Jameson's behaviour that they considered to be well founded. In particular, they felt that 'very serious evils have resulted from the extraordinary difficulties experienced by men of Science in obtaining access to the Museum, and the general prohibition enforced against the fair use of it'.[61] Before the Commission he was quite unrepentant on the subject. Asked by the Commissioner 'Do you think it expedient the Keeper should be possessed of such power of withholding the use of the Museum from Scientific persons, or giving or refusing them permission to make drawings, according to his option?', he answered: 'I think it a beneficial check. I have prevented the publication of works which would have done no credit to Science.'[62] That Jameson could be a difficult character is without doubt. He even seems to have alienated some important figures by his manner and behaviour with whom earlier in his career he had enjoyed a close relationship. In his *Memoir* of the life of John Fleming, John Duns quotes Fleming, the influential natural historian and Church of Scotland minister, as complaining, probably sometime in the mid-1820s, of his treatment by Jameson and comparing him unfavourably with David Brewster, with whom Jameson had shared the editorship of the *Edinburgh Philosophical Journal* until 1826:

> I have found Dr B.'s [Brewster's] friendship uniform, and kind, and intimate – 'the council's' [Jameson's] irregular, cold, and distant. As men of science there can be no competition. 'The council' is bolstered by his professorial chair and the museum. Dr B. stands on a broad foundation of discovery and generalisation. Dr B. has mentioned my name on suitable occasions with respect; the Prof. has erased mine from his editions because it was coupled with *Thomson's Annals of Phil.*[63]

In the early 1830s there emerged a challenge to Jameson's authority from an unexpected source. The attack came from a medical student at the University, Henry H. Cheek (1807–33), who was also one of the two editors of the short-lived *Edinburgh Journal of Natural and Geographical Science* (1829–31). Cheek, clearly a very disgruntled student, used his journal to launch a campaign against Jameson. He first attacked him in his role as president of the Wernerian Natural History Society. The Society counted among its members most of the important figures in natural history in Scotland in the first half of the nineteenth century. However, according to Cheek, who was not himself a member, all was not well at the Society. In May 1830 the editors of the *Edinburgh Journal of Natural and Geographical Science* congratulated themselves on 'having instigated the present investigation of the independent members of the Wernerian Society into the singular condition of their mis-directed institution'.[64] A further tirade against the direction of the Society followed in the July number. Among a catalogue of complaints that Cheek directed at the Society, the main issue around which the dispute crystallised was the lack of access of members to the Society's library, which, he claimed, appeared to be reserved for the sole use of Jameson. There followed a very frank exchange of views between Cheek and Patrick Neill (1776–1851), the secretary of the Wernerian Society, who came to Jameson's defence. This controversy was conducted largely through a series of journal articles and pamphlets through the pages of which it is possible to chart the course of the dispute. The tone of the exchange rapidly became very personal. In response to Neill's first reply to his journal article in an 'Address to the Members of the Wernerian Natural History Society', Cheek attacked Jameson personally in his position as professor of natural history and keeper of the Natural History Museum:

> I can declare that, during the four years of my residence in Edinburgh, I have been grieved to see the Museum of the University closed to the student who did not purchase certain nominal privileges at an exorbitant price, and, what was more disgraceful, the total uselessness of that establishment to the man of science; – I have felt indignant at the perusal of the syllabus of lectures which the Professor of Natural History puts into the hands of his pupils, and which is only calculated to delude; and I have beheld with disgust a coterie brooding like a night-mare over the Wernerian Natural History Society, till there was little remaining of it but the mockery cast by its name, upon opinions which are now only to be found in the pages of the history of error.[65]

In the 'miscellaneous information' section of the December 1830 number of the journal there then appeared a paragraph by Cheek noting that the Royal Commission for Visiting Scottish Universities had just given its report; the editors announced that they 'look anxiously for the judgment which may be passed upon those flagrant malpractices which we have already exposed', presumably a reference to Jameson's conduct over the Museum, which was indeed roundly criticised in the report of the Commission.[66] Cheek also had a personal interest in the report, as he had himself successfully petitioned to give evidence to the Commission, which he did on 25 January 1830.[67] In this testimony he was extremely critical of Jameson's management of the museum. Among numerous other complaints, he claimed that Jameson had denied him access to some specimens of the marine invertebrates that Robert Grant had added to the collection. In his second pamphlet, Neil accused Cheek of 'doing all in his power (fortunately little) to hold up to contempt and infamy either its President or Secretary, or both, by the grossest imputations.'[68] He also accused Robert Knox, who had supported Cheek, of gross ingratitude to Jameson, who had helped him in his early career.

The passage quoted above suggests that, from Cheek's perspective at least, there were three main complaints regarding Jameson's conduct as professor of natural history. First, was the issue of Jameson's management of the Museum, to which he restricted access in an arbitrary fashion. The justice of this complaint would seem to be confirmed by the report of the Scottish Universities Commission. The second concerns Jameson's teaching, which it is implied did not live up to the promise of the published syllabus. Unfortunately Cheek does not provide any more detailed evidence to back up this latter assertion, and it is not confirmed by other sources. Third, Cheek claims that Jameson and his cronies mismanaged the affairs of the Wernerian Society, and, in particular, that he treated the valuable library of the Society as his personal property. He also claimed that Jameson poached papers read at the Society's meetings for his *Edinburgh New Philosophical Journal*, thus depriving the *Memoirs* of the Society of these papers. It is certainly true that the *Memoirs* were only published erratically and at increasingly long intervals, only running to eight volumes between 1808 and 1838.

Taken as a whole, the dispute gives us an invaluable insight into the ambiguous figure cut by Jameson in the first years of the 1830s. Given the extremely negative accounts of Jameson broadcast by Cheek, and to a lesser extent by Darwin and Fleming, it would be possible to speculate that those who praised him did so purely out of self-interest. However, this does not ring entirely true. The members of the Universities Commission found good reason to praise his teaching while castigating him for his management of the museum. The warm, positive sentiments expressed by many of his former students decades later, when they would have had little to lose by criticising him, must also be taken into account. Viewed in conjunction with the other evidence available, what emerges from the various, and often apparently contradictory, accounts of Jameson's character and professional conduct is the picture of a hard-working and dedicated professor who devoted his life to his subject and his students, but who could at times be cantankerous and jealous of his own prerogatives.

## COMPARATIVE ANATOMY AT THE EXTRA-MURAL MEDICAL SCHOOLS

The University was not the only place where students could learn about the latest developments in natural history. Many of the extra-mural medical schools that existed in Edinburgh in the early nineteenth century also provided comprehensive courses in comparative anatomy. John Barclay was the proprietor of the most successful of these. Barclay's MD thesis had been entitled *De Anima, seu Principio Vitali*, and the existence of a vital principle was an abiding interest for him throughout his career.[69] In a book he published on anatomical nomenclature he explained the vital principle as follows: 'In every living organised structure, there is plainly a power that preserves regulates, and controls the whole.'[70] This principle was an immaterial agent the action of which could not be explained by the normal laws of chemistry or physics; it was 'independent of either configuration or matter, extends its influence over the nutritious particles around it, as fire does over combustible materials, and which assimilates these particles in a regular manner'.[71] In a letter dated 31 December 1821 from the German phrenologist Johann Gaspar Spurzheim (1776–1832) to his Edinburgh disciple George Combe (1788–1858), Spurzheim noted that Barclay's hostility to phrenology was a result of his attachment to the idea of the vital principle, concluding that 'Dr Barclay has not sufficient reason to believe in Phrenology, since it does not agree with his vivifying principle, which builds the organisation'.[72]

The existence of a vital principle was an abiding concern for Barclay throughout his career, leading him to adopt a strongly anti-materialist stance. His hostility towards materialism was reinforced by his strong religious beliefs. His last published work, *An Enquiry into the Opinions, Ancient and Modern, Concerning Life and Organization* (1822), consisted of a sustained attack on a rogues' gallery of thinkers whose views Barclay disagreed with, because of either their materialism or their unsatisfactory views on the vital principle, from Epicurus and Aristotle to Buffon and Erasmus Darwin.[73] In the preface to his book, Barclay also stated that it was intended as an antidote to objectionable ideas for impressionable medical students: 'As young men entering on the studies of anatomy and chemistry, are naturally let to investigate the causes of organisation, and are frequently apt to form hypotheses on grounds of information not the most ample, it was imagined that such an inquiry might be particularly useful to them.'[74] This book provides a valuable insight into both Barclay's own reading and the theories that were likely to have been discussed in medical circles in Edinburgh in the 1820s.

There is surprisingly little evidence of the views of Barclay's successor Robert Knox in the 1820s. His publications during this period were generally rigorously factual, and give very little indication of his theoretical stance. Knox himself destroyed the bulk of his papers before his death, and the material utilised by his biographer Henry Lonsdale and Knox's correspondence with bodies such as the Edinburgh Royal College of Surgeons have been lost, depriving the historian of potentially valuable sources of information on his early career.[75] His adherence to Geoffroy's philosophical anatomy and rejection of Cuvier's functionalist approach, which assumed that form is perfectly adapted to function, is hinted at in a few of the publications from his Edinburgh years, for example in a paper he gave in 1826 on the vestigial spur found in the foot of the female Echidna he noted: 'The physiological anatomist can have no difficulty in comprehending that this organ must bear to the male spur the same relation that the human male breast does to the female.'[76] Otherwise his opinions on comparative anatomy and evolutionary thinkers in the 1820s have to be reconstructed from biographical sources and writings from later decades, when Knox was living in London. The potential dangers of such an approach can be illustrated by comparing his stated views on the effect of climate on the human species in 1824 with the opinion he gave on the same subject in 1850. In 1824, he had this to say on the origins of the different races:

> We may view the human race as derived originally from one stock, to which the arbitrary name of Caucasian has been given. This species, influenced by climate and civilization, assumed, at a very early period, five distinct forms, which have also been arbitrarily designated by the names Caucasian, Mongolic, Ethiopian, American, and Malay.[77]

By 1850, however, he considered that never 'has climate or external circumstances effected any serious changes, produced any new species, any new groups of animal or vegetable life, any new varieties of mankind'.[78] It seems that between 1824 and 1850 Knox had completely reversed his opinion on the role of conditions of life in generating races. It is quite possible that other opinions, undocumented in his writings of the 1820s, had undergone similar alterations in the intervening decades. Where these opinions appear to be derived from ideas found in Geoffroy's writings of the 1820s, we might be on slightly safer ground. It is certainly the case that his continuing adherence to the doctrines of transcendental anatomy and unity of plan he had learned from Geoffroy in Paris represented an area of evident continuity in Knox's thought. Nevertheless, the danger of assuming that Knox's views had remained unchanged during a period of twenty-five years cannot be overstated.

Another important figure who taught at Barclay's school was Robert Grant, who completed his medical studies at the University of Edinburgh in 1814. Grant was first introduced to transformism through the writings of Erasmus Darwin before he encountered the theories of Lamarck. He seems to have been aware of Darwin's transformist theories at least since his years as an undergraduate at the University of Edinburgh, as he made reference to Darwin's *Zoonomia* in his undergraduate dissertation.[79] Much later, in the dedication to Charles Darwin in Grant's *Tabular View of the*

*Primary Divisions of the Animal Kingdom* (1861), he wrote that '[m]ore than fifty years have now elapsed since the "Zoonomia" of your illustrious ancestor, Dr. Erasmus Darwin, first opened my mind to some of "the laws of organic life"'.[80] An inheritance from his father, who died in 1808, had left him at liberty to spend the next decade or so studying and travelling extensively in Western Europe. In 1815 Grant was in Paris, studying

**Figure 3.4** Lithograph of Robert Grant by T. Bridgford. Credit: Wellcome Collection. CC BY.

under Henri de Blainville, from where his studies took him to Italy and Germany. In 1820 he was back in Edinburgh, where he practised medicine for a while and attended the natural history classes of Robert Jameson in 1823, whose lectures he had first experienced as a medical student in 1810.[81] In the same year as his return to Edinburgh he became a member of the Linnaean Society, helped by a recommendation from Jameson, and was elected a member of the Royal Society of Edinburgh in 1824.[82] He also became a member of the Wernerian Natural History Society and an honorary member of the Plinian Natural History Society.[83] During the 1820s he visited Paris regularly in the summer and was personally acquainted with many of the major figures of French natural history, including Geoffroy and Cuvier. He was also a friend of John Fleming, who named a genus of sponge *Grantia* after him, 'to commemorate his valuable services in elucidating the physiology of sponges'.[84]

Grant had attended John Barclay's lectures on comparative anatomy in 1821,[85] and in 1824 he was invited by Barclay to teach the part of his course that dealt with invertebrates, which was Grant's particular area of expertise. It was in relation to these simple creatures that he was principally to develop his own ideas regarding the transmutation of species. In 1827 Grant left Edinburgh to take up the chair of zoology at University College London, a post for which he was recommended by an impressive list of Scottish scientific figures from both inside and outside the University of Edinburgh, including Robert Jameson, David Brewster, John Fleming and Alexander Monro, tertius. Sadly, beyond his published work, very few sources exist for Grant. Not only do we know that he was in the habit of burning manuscripts of his work, but also very few of his letters seem to have survived, although we do know that he kept bound volumes of his correspondence during his lifetime.[86] These volumes should have been bequeathed to University College London on his death, but do not seem to have ever come into the possession of the University.[87] As a result, the outlines of his life and thought have to be reconstructed largely from published sources.

We know little of what Knox and Grant actually taught at John Barclay's school, as no evidence in the form of student notes or published syllabuses have survived, and their lectures were never published. One extra-mural lecturer whose course we know a great deal more about, because he published a book based on his lectures, is John Fletcher (1792–1836). Fletcher graduated from Edinburgh in 1816 and taught at the Argyle Square Medical School from 1828 to 1836. He was an admirer of Geoffroy and an active promoter of unity of plan and recapitulation theory. His lectures were published as *Rudiments of Physiology* (1835–7), in what was 'the first book completed on higher anatomy in Britain'.[88] He frequently cited Grant in his book and was clearly also aware of the ideas of Knox. Although he was an advocate of the new philosophical anatomy, it becomes quickly apparent on reading *Rudiments of Physiology* that he was no transformist. He did, however, discuss evolutionary theories at some length in this work. An exposition and critique of these theories therefore certainly formed part of the syllabus that Fletcher taught his students.

## NATURAL HISTORY, SCIENTIFIC AND MEDICAL SOCIETIES

While visiting Paris in the late 1820s, a recent graduate of Edinburgh University and member of the Plinian Natural History Society by the name of John Coldstream was afflicted by a severe moral and religious crisis. William Mackenzie, a Scottish doctor who knew Coldstream in Paris, explained his breakdown in the following manner: 'Though a young man, I believe, of blameless life, still he was more or less in the dark on the vital question of religion, and was troubled with doubts arising from certain Materialist views, which are, alas! too common among Medical students.'[89] While a medical student at Edinburgh, Coldstream had been an active member of the Plinian Natural History Society, which he had joined on 18 March 1823. Charles Darwin knew him in Edinburgh and described him as 'prim, formal, highly religious and most kind-hearted'.[90] He was also a close associate of Robert Grant and was one of a group of students who regularly accompanied him on natural history collecting trips to the Firth of Forth. While Coldstream steered away from controversial subjects in his own contributions to the Society, it has been suggested that his Paris crisis had something to do with the materialist ideas current among students that Mackenzie referred to, ideas for which the Plinian Society provided a forum for discussion, and to which Coldstream would have been exposed at its meetings.[91]

The Plinian Natural History Society, named after the great Roman natural historian Pliny the Elder (AD 23–79), had been founded in 1823 by a group of undergraduates at the University of Edinburgh. It has sometimes been incorrectly stated that it was founded by Robert Jameson. This misapprehension can be traced back to Darwin, who suggested that Jameson was the founder in his 'Recollections',[92] and this has subsequently been repeated as fact by a number of scholars, including Desmond and Moore in their biography of Darwin.[93] Although Jameson was the 'senior honorary member' of the Society, he never actually attended any of its meetings. When asked about the Plinian and his relationship with it by the commissioners of the Royal Commission on the Scottish Universities in 1826, he described it as a forum 'where young men can discuss subjects which they have heard in the class of Natural History, or in the other classes where such subjects are considered'.[94] While he stated that the society had been founded 'under his countenance', he denied that he had 'any particular controul [sic]' over it. However, he went on to affirm that in his opinion the members 'distinctly confine themselves to the proper subjects of investigation'.[95] As most, if not all, members would have been Jameson's students, his teaching undoubtedly had a significant influence on the debates that took place at the Society's meetings.

According to a biography of John Baird, one of the founders of the society, Baird and his associates, 'feeling the want of a society where younger students of nature could meet and discuss their views freely among themselves, unawed by the older and more mature naturalists of the day, resolved to institute a society for the advancement of the study of natural history, antiquities, and the physical sciences in general'.[96] Foremost among the founders of the society was Baird, along with his brothers Andrew and William. Both William Baird's *Memoir of the Late Rev. John Baird* (1862) and a *Nature* article on 'Local scientific societies' published in 1873 credited John Baird as the founder of the society.[97] According to the society's minutes, it was also Baird who announced the 'proposed plan and objectives' of the society at its first meeting on 14 January 1823.[98]

By June 1826 its membership had risen steadily to 106. It also had twenty-five corresponding members and ten honorary members. The honorary members were Edinburgh University professors Robert Jameson (natural history), Thomas Hope (chemistry), Robert Graham (botany), Andrew Duncan senior (institutions of medicine) and Andrew Duncan junior (materia medica), George Husband Baird, the Principal of the University, and from outside the University John Fleming, Sir Walter Scott, Sir Humphry Davy and Sir James Hall. In 1829, the *Magazine of Natural History* included the following description of the objectives of the Society:

> The principal intention of the founders of the Plinian society was to promote natural history; but antiquarian researches, and the advancement of all the physical sciences, have also been included among its professed objects. The means which have been adopted for the prosecution of the views of the Society are, the reading of papers, debates, the formation of a museum and library, and excursions to the country, for the examination and collection of objects of natural history. Papers have been read on subjects connected with all the departments of natural science, more especially on the zoology, botany, geology, mineralogy, meteorology, and antiquities of Scotland.[99]

In June 1826 the minutes of the Society noted that a deputation to Professor Jameson could report on 'the very flourishing condition of the Society – that there are at present about 150 members on its list and that although we have no compulsory laws and impose no fines, we have had a good attendance of members all session and an abundant supply of papers'.[100] The Society continued to flourish and attract new members throughout the 1820s, its weekly meetings between November and July regularly attracting around twenty members. As the outline of the Society's activities above indicates, papers given at the society addressed a very wide range of topics, from mineralogy and ornithology to more controversial topics, such as phrenology, the existence of a vital principle, spontaneous generation, and the origin and transmutation of species; papers were even read to the society on subjects such as apparitions and astrology. In 1829 the *Magazine of Natural History* could still report that: 'It will be unnecessary to remark on the flourishing state of this Society, when it is stated that it is at present composed of upward of 180 members'.[101]

The Society, however, went into a marked decline after 1830. The minutes for 6 March 1832 refer to its 'declining state',[102] and in those for the next meeting on 3 April the Convenor of the Committee, Henry Cheek, reported that the Committee considered that 'the Plinian Society is not in a condition to be continued any longer as a separate institution', a situation that he ascribed to the 'apparent apathy in this University to the cultivation of those Branches of natural science for the prosecution of which this Society was originally founded'.[103] The society did, however, continue to meet at increasingly irregular intervals until 19 May 1835, after which there was a lapse of almost six years until January 1841, when the minutes note that since 'there has been no meeting of the Society since 1835 and there was no prospect of any revival of it taking place, it was moved by Dr Spittal that the Society be dissolved'.[104]

The evidence of the Society's minutes sheds much light on the prevalence among its members of the kind of materialist ideas that seem to have shocked Coldstream so much; for example, in a famous and widely discussed incident, the account of a paper given at the Plinian Society by William A. F. Browne (1805–85) on 27 March 1827 was struck from the minutes of the Society by an unknown hand.[105] The announcement that the paper was to be given was also deleted from the minutes of the previous week's meeting. If the intention of the unknown person was to erase all trace of the paper from the minutes, he did his work badly, for the account of the paper is still clearly legible. Curiously, the following week's minutes record that those from the previous week's meeting had been approved, with no suggestion of any debate or controversy regarding Browne's paper.

Unfortunately, in the absence of further evidence, no firm conclusion can be drawn regarding the circumstances of the deletion.

In 1827 Browne had recently qualified as a licentiate of the Royal College of Surgeons of Edinburgh.[106] He was also a member of the Phrenological Society and a friend and follower of George Combe. Combe was the most influential phrenologist in Britain in the middle decades of the nineteenth century. Central to phrenology, a would-be new science of the mind, was the belief that the brain was composed of a number of separate organs, each of which corresponded to a faculty of the human mind. Phrenologists believed that the shape of the skull faithfully mirrored the form of the brain within in such a way that the development of these different organs could be read by a skilled practitioner from the shape of the skull. Its enemies tried to associate phrenology with materialism, although Combe himself was very careful to dissociate himself from any such suggestion. At a Phrenological Society dinner in 1829, Combe described Browne as 'a member of the Society, who on all occasions had displayed much zeal in the cause of Phrenology, and who at the present time had been particularly active and successful in bringing before the medical students of Edinburgh the importance of Phrenology as the doctrine of the functions of the brain'.[107] As the superintendent of first Montrose Asylum and then Crichton Royal Asylum in Dumfries, Browne was to go on to have a distinguished career as a bold innovator in the treatment of mental illness. It may have been his interest in phrenology that inspired Browne's controversial paper. The deleted account of Brown's paper expresses just the kind of materialist opinions that phrenology's critics feared were a natural consequence of its doctrines.

The account of Browne's paper in the Society's minutes is as follows:

Mr Browne then read his paper on organization as connected with life & mind in which he endeavoured to exhibit the following propositions.

I. That all matter is organized.

II. That it is the gradually increased perfection in the arrangement of the parts constituting organization which is the source of the distinction perceptible in the various objects of nature & not specific differences.

III. That life is the abstract of the qualities inherent in these modes of arranging matter.

IV. That mind is to be distinguished from life, being neither one of the functions or combination of qualities, by the concatenation of which life is constituted nor a term indicating a similar idea.

And V That mind as far as one individual sense & consciousness are concerned, is material.

A discussion ensued between Messrs Binns, Greg, Mr Grant, Ainsworth & Browne.[108]

Although the account of it is somewhat cryptically worded in places, Browne's March 1827 paper appears to have been strikingly materialist in essence. That Browne considered the human mind to be a product of the physical organisation of the brain is unquestionable on the evidence of the fifth point. In harmony with his attempts to banish immaterial agents from explanations of mental phenomena, Browne had given a paper on 25 April 1826 to the Society that presented a materialist explanation of apparitions. In it he attempted to prove 'that, there exists in the mind a primitive feeling of belief in spectral appearances, that as a superabundant circulation of the blood in some parts of the brain induce a state of mind calculated for receiving these'.[109] Thus an immaterial, spiritual phenomenon could be reduced to the manifestation of a purely physical process within the brain. His explanation of the phenomenon of life as 'the abstract of the qualities inherent in these modes of arranging matter' is perhaps less clearly stated than his views on the mind. However, the second point seems to imply that the differences between living and non-living things are the result of greater or lesser 'perfection' of organisation, rather than being evidence of a qualitative, 'specific' difference. In the light of this, it seems fair to conclude that Browne also believed that life was also not a manifestation of an immaterial 'vital principle', but rather a product of organisation. Such radical materialism was perhaps not surprising from an enthusiastic young disciple of phrenology, but would have been deeply shocking to the more conservative among Browne's fellow students, probably including the 'highly religious' John Coldstream. However, Browne does not seem to have been alone in attributing life and mind to material causes. On 25 March 1828 Thomas Shapter presented a paper to the Society in which 'he stated his opinion to be that vital principle was the result of organization'.[110] It was perhaps to combat such views that John Symonds presented a copy of John Barclay's fierce defence of the vital principle, *An Inquiry Into the Opinions, Ancient and Modern, Concerning Life and Organization*, to the society on 19 June 1827.[111] Based on the evidence of the Society's minutes, it is clear that a number of members flirted with materialist theories of the nature of organic life and the human mind, and that such ideas were openly discussed at the Society's meetings, even if some members were clearly actively hostile to them, as we can see from the case of Browne's February 1827 paper. It should perhaps not be surprising then that, as we will see in subsequent chapters, a number of members also touched on evolutionary themes in the meetings of the Society.

Like the Plinian Society, the Royal Medical Society was founded in 1737 by and for medical students at the University of Edinburgh. Most members of the Plinian Society in the early nineteenth century, including Darwin, were also members of the Royal Medical Society. However, unlike the Plinian Society, this Society kept books of all the dissertations that had been read to the Society by its members. These provide a great deal more detail than the frustratingly cursory accounts of papers given in the Plinian Society's minutes and proceedings. The rules for providing subjects and delivering papers to the Royal Medical Society were laid down in the code of laws in 1781. According to these rules, each ordinary member had to provide 'the History of a Case, a Medical or Philosophical Question, and an Aphorism of Hippocrates'.[112] The committee then selected thirty-six of these as the subjects for dissertations. The rules then laid down that the president would order 'each member who intended to write during the next session to declare his intention at the next meeting, when each of the thirty-six oldest members who intended to write was required, from numbers on slips of paper, corresponding with those marked on the sets, to draw one'.[113] In 1796 the aphorism was dropped, as was the case study in 1816, leaving only the medical or philosophical question to be answered in the dissertation. In practice the system was not as rigid as these rules would suggest, and students were very often allowed to write on subjects of their own choosing.[114]

While the bulk of dissertations presented to the society were on medical subjects, a substantial minority were on natural history topics. These included papers on the debate between Wernerians and Huttonians in geology, theories of generation, the relationship between mind and organisation and the origins and nature of the human races. The latter was one of the more popular non-medical subjects for dissertations. Six dissertations were presented to the Society on the subject between 1800 and 1836. Out of these, five supported the monogenist view that all the races had a common origin, against only one polygenist essay. As we will see in subsequent chapters, one of these dissertations on race, by Henry H. Cheek, also presented the outline of a theory of the transmutation of species. In general, the non-medical topics dealt with are similar to many of those found in the minutes of the Plinian Society.

The Royal Physical Society, was founded in 1771 as the Physico-Chirurgical Society, before changing its name in 1778. Like the Royal Medical Society it provided a forum for the debate between monogenists and polygenists over race, which can be traced through its surviving books of dissertations. The only explicit reference to transformism as such is in a dissertation given by Ralph Smyth Stewart on 18 October 1822. In his paper on theories of the earth, he gave a somewhat mocking account of the transformist theories of the French philosopher Benoît de Maillet (1656–1738), to which he added that:

> Many bold imaginations have followed up this opinion, and have ascribed the existence of animals to the universal fluid, which were at first of the simplest kind, as the monads and other infusory microscopic animals: that in process of time these animals became complicated and assumed that diversity of nature and character in which they now exist.[115]

Stewart himself considered it 'strange that these systems which are apparently as wild as those of the illustrious philosophers of Laputa, have been generally conceived by men of science and genius'.[116] Although Stewart does not name the men of science he is referring too, it is evident that he was aware of the transformist theories in circulation at the time, Erasmus Darwin or Lamarck being likely candidates as the targets of his mockery. While negative in tone, Stewart's remarks certainly demonstrate the currency of transformist ideas among the members of Edinburgh's scientific and natural history societies at this time.

Edinburgh's other natural history society in this period, the Wernerian Natural History Society, was quite a different beast from the Plinian. Membership was restricted to graduates and was not open to students. It counted among its members most of the leading Scottish natural historians of the day. The Society was founded on 12 January 1808 by Robert Jameson, who was also its president until his death in 1854. Its other founder members were William Wright, Thomas Macknight, John Barclay, Thomas Thomson, Colonel Stewart Murray Fullerton, Charles Anderson, Patrick Neil and Patrick Walker. These nine *'resolved to associate themselves into a society for the purpose of promoting the study of natural history*; and in honour of the illustrious Werner of Freyberg [sic], to assume the name of the Wernerian Natural History Society'.[117] Jameson was the most important of Werner's disciples in Britain. It is clear that one major aim of the society in its early days was to combat the theory of the earth proposed by James Hutton, and popularised in John Playfair's *Illustrations of the Huttonian Theory of the Earth* (1802).[118] Eight volumes of the *Memoirs of the Wernerian Natural History Society* were published between 1811 and 1839. As its name would suggest, the focus of the society was on mineralogy and geology, although the *Memoirs* also contain a significant proportion of papers on zoological and botanical subjects. Active members during the late 1820s and 1830s included

Robert Grant, Robert Knox, John Fleming and William MacGillivray, all of whom regularly contributed papers to the *Memoirs* of the Society during its heyday in the 1820s. The Society went into a decline, along with the health of its president, in the 1840s, and was finally formally wound up in 1858, four years after Jameson's death.

## NATURAL HISTORY AND SCIENCE JOURNALS

Three science and natural history journals important for our story were published in Edinburgh in the first half of the nineteenth century. Along with David Brewster, Jameson co-edited the *Edinburgh Philosophical Journal* from 1819 until 1824, when Brewster left, leaving Jameson in sole charge of the journal. From 1826 it was renamed the *Edinburgh New Philosophical Journal*, with Jameson still at the helm.[119] Under this new guise it continued publication until 1852. The *Edinburgh Journal of Science* was edited by David Brewster after he stopped working with Jameson on the *Edinburgh Philosophical Journal*.[120] The third journal was the *Edinburgh Journal of Natural and Geographical Science*, edited by Henry H. Cheek and William Francis Ainsworth, both medical students at the university at that time. The most important of the three in terms of the number and variety of the evolutionary articles it published was the *Edinburgh New Philosophical Journal*, so it is with this periodical that I will begin.

The *Edinburgh New Philosophical* Journal is notable for publishing a significant number of evolutionary articles in the late 1820s. This journal was one of the most important periodicals in its field and attracted contributions from many significant figures in natural history from across Europe and beyond. The full original title of the journal was *The Edinburgh Philosophical Journal Exhibiting a View of the Progress of Discovery in Natural Philosophy, Chemistry, Natural History, Practical Mechanics, Geography, Navigation, Statistics, and the Fine and Useful Arts*, which gives an accurate sense of its scope. It generally included around twenty-five articles on extremely diverse subjects, followed by the proceedings of major Scottish scientific societies, then a section of 'scientific intelligence' under various subject headings, and finally lists of patents. The journal published not only articles by important figures in Edinburgh natural history circles, such as Robert Grant, Robert Knox, John Fleming, William MacGillivray and Robert Graham, but by major national and international figures such as Alexander von Humboldt, Georges Cuvier and John Audubon. To give a flavour of the eclectic nature of the journal, volume 8, number 25 (1825) contained the following among its articles: 'On the construction of oil and gas burners', by Robert Christison, professor of medical jurisprudence at the University of Edinburgh; 'A table of the geographical positions of several places in India', by James Franklin; 'Observations and experiments on the structure and function of the sponge', by Robert Edmond Grant; and 'Sketch of the geology of Sicily' by Charles Daubeny, professor of chemistry at the University of Oxford. A significant proportion of authors were current or former students of Jameson's natural history class. After Jameson became the sole editor in 1824, it was widely referred to in natural history circles simply as 'Jameson's journal'.[121] Apart from the slight change to the title after 1826, the editorial policy of the journal seems to have been largely unaffected by Brewster's withdrawal from the editorship. Under Jameson's sole direction it was notable for the appearance of a number of evolutionary articles. The sudden appearance of articles dealing with the transformation of species might suggest that Jameson had a more sympathetic attitude to evolutionary ideas than Brewster, who was severely evangelical in his religious views and became a great opponent of evolution in the aftermath of the publication of Robert Chambers' *Vestiges of the Natural History of Creation* the 1840s, but it is impossible to know for sure. The editorship of the *Edinburgh New Philosophical Journal* gave Jameson considerable power of patronage in Edinburgh. He used this to provide a platform for many of his younger associates, including Grant and Knox, to publish their ideas. It also allowed Jameson to promote theories in the natural science that he favoured, such as the work of the Wernerian geologists Jens Esmark (1763–1839) and Ami Boué (1794–1881).

After he relinquished the editorship of the *Edinburgh Philosophical Journal*, David Brewster established his own journal, *The Edinburgh Journal of Science*, in 1824, which he continued to publish until 1832. Brewster's editorial policy in his new journal was very close to that of the *Edinburgh Philosophical Journal* and it published a similar range of articles. According to Brewster, the 'same kind of information which has been given in the Edinburgh Philosophical Journal, will be given in the present one, with various improvements, both in the embellishments, and in the scientific character of the work'.[122] It seems therefore that Brewster saw his journal as a rival to Jameson's journal and his decision to establish it a direct result of the breakdown of his editorial relationship with Jameson. Editorial assistance was provided in a number of areas by a group of experts, including Robert Knox in zoology and comparative anatomy, and Samuel Hibbert in geology and antiquities.[123] Some writers, including both Knox and Grant, continued to publish articles in both Brewster's and Jameson's journals.

Between October 1829 and June 1831 Henry H. Cheek edited the *Edinburgh Journal of Natural and Geographical Science*. For the first two volumes his co-editor was William Francis Ainsworth, his friend, fellow medical student and member of the Plinian Society. The final volume was edited by Cheek alone, while editorial assistance in specific areas provided by the naturalists Sir William Jardine (1800–74) and John Scouler (1804–71), the botanist G. A. Walker Arnott (1799–1868), the agricultural chemist and mineralogist J. F. W. Johnston (1796–1855) and the anatomist Robert Knox. Cheek had therefore succeeded in mustering a number of significant figures in Scottish science and natural history to help him, who had 'undertaken the entire direction of their several Departments'.[124] In total twenty-one monthly issues of the journal were produced. The *Journal* was published in Edinburgh but also distributed in London and Dublin.[125] It seems that Cheek may have published this journal from his own house in Gardner's Crescent, Edinburgh, as Patrick Neill, the Edinburgh printer and secretary of the Wernerian Natural History society, referred to 'his converting his dwelling-house into a printing office' in a polemical pamphlet he wrote attacking Cheek in 1830.[126]

The journal published articles on a wide variety of scientific and geographical subjects, including some by notable figures in Edinburgh natural history circles, such as William MacGillivray, William Jardine and John Fleming. Despite the youth and inexperience of its founders, it therefore seems to have had the support of a number of important figures in Edinburgh natural history. Although monthly rather than quarterly, it was similar in scope and format to Jameson's *Edinburgh New Philosophical Journal* and may have been conceived as a competitor to it. As we have already seen, Cheek had used the pages of his journal to openly criticise Jameson in no uncertain terms, prompting a brief pamphlet war with Patrick Neill, who accused Cheek of trying to set himself up as 'the Baron Cuvier of Edinburgh'.[127] As we will see in later chapters, Cheek seems to have used the journal not only to attack those he perceived as his enemies, but as a platform for his own evolutionary views.

Cheek's bitter attacks on Neill and Jameson must surely have irreparably soured relations with many of the leading figures in the natural history establishment in Edinburgh. The dispute gives us an invaluable window into Cheek's position in the world of Edinburgh natural history in the first years of the 1830s. It seems that Cheek was at the centre of a circle of younger natural historians who met for regular *conversaziones* at his house in Gardner's Crescent. From this position he seems to have felt confident enough, perhaps unwisely, to challenge the authority of the Wernerian Society, and of its president, Robert Jameson. It might be speculated that this controversy, in which Neill seems to have had the last word, may have led, directly or indirectly, to the winding up of Cheek's journal after its next volume and hastened his departure from Edinburgh after his graduation in 1832, but it is impossible to know for certain.[128]

In any case, the story does not end well for Cheek. In 1842 William Francis Ainsworth wrote an appreciation of his friend in a footnote to an account of a descent he and Cheek had undertaken into Eldon Hole cave in Derbyshire some years before in a journal he edited. Cheek had died in 1833, and Ainsworth gives a poignant tribute to his friend and an account of his early death, it would seem by suicide. If Ainsworth is to be believed, his dispute with Neill and Jameson had blighted his subsequent career on his return to his native Manchester and led directly to his decision to take his own life. This is what Ainsworth has to say about the character of his friend and the circumstances leading to his tragically early death:

> He was a young man of gentlemanlike feeling, good classical acquirements, and most zealous talent, but sanguine, and nervously susceptible. This temperament led him to hold the pettifogging of the world in great abhorrence, and brought from his pen a series of most caustic exposures of the Wernerian Society and the University Museum. The same peculiarity of mind had led him to hope that, labouring in the honourable career of science, he was reaping distinction for himself in after life; but when, after taking out his medical diploma, he went to establish himself in his profession at Manchester, he found that to meddle with science was to be expelled from all fraternity in the profession; the effect upon his mental constitution was so great, that he never recovered the shock, and shortly afterwards withdrew himself – by a sad, self-inflicted death – from a world he appears to have been unfitted for.[129]

The role played in Scottish science by Robert Jameson, Edinburgh's professor of natural history for half a century, cannot be over-emphasised. As we have seen, Jameson's influence could be felt in every corner of the scientific scene in Scotland. His role as professor of natural history allowed him to influence generations of students, some of whom would go on to revolutionise thinking on the natural world in Britain in the decades to come. In the lecture theatre, on field trips, in discussions with his students in the natural history museum, his enthusiasm for his subject was infectious, even if his lecturing style could be a little dry. Jameson took an active interest in the latest ideas on the natural world and among his students there was a veritable ferment of intellectual activity that found an outlet in some of the student societies, where exciting new continental developments in philosophy and science were discussed and debated. As the founder and president of the Wernerian Society he brought together a glittering company of Scotland's leading natural historians to share ideas and theories. Although he had no formal connection with the Plinian Society, it was formed 'under his countenance' and, as most of its members were his students, his opinions must have loomed large over its deliberations. His editorship of the *Edinburgh Philosophical Journal* and then the *Edinburgh New Philosophical Journal* gave him a platform for the dissemination and promotion of theories that were in sympathy with his own views. The combination of all these roles gave him a tremendous power of patronage in Scottish natural history circles. He used this power to promote the careers of some of the most radical thinkers of their generation. In particular, Robert Grant and Robert Knox owed a great deal to Jameson. He was not without flaws, and it seems that Jameson had little time for those who did not share his enthusiasm or did not agree with his views. We have the testimony of Henry Cheek, Darwin and Fleming to show that he was by no mean universally loved and respected, and it is clear that he could be a difficult character to deal with. He also seems to have turned a number of important friends and allies against him in later decades.

Although Jameson, along with a group of his fellow professors, had actively blocked John Barclay from gaining a chair at the University, he played a major role in promoting the career of Barclay's successor Robert Knox, as well as that of Robert Grant, who also taught at Barclay's school. Infused with the new French transcendental anatomy, Knox and Grant played essential roles in propagating radical new ideas among the medical students of the University. One of the students whom Knox almost certainly inspired was Henry Cheek. Cheek, although no friend of Jameson, added to the mix through the publication of the *Edinburgh Journal of Natural and Geographical Science*, which actively promoted new and controversial ideas on the nature and origin of species. Like Jameson, Cheek published the work of both the leading figures in Scottish natural history and the exciting new ideas coming from continental sources that were revolutionising discourse on the natural world in those years. The soirées that he hosted, and to which Neill's hostile testimony bears witness, must likewise have been the venue for the airing of a heady blend of new ideas in the natural sciences, although sadly they have left little trace in the written record. In Chapter 5 I will be looking in more detail at the brief career of Henry Cheek as an evolutionary thinker and the reception and influence of French comparative anatomy and evolutionary theories in Edinburgh more generally. But first, I need to look at the geological theories espoused by Jameson and his circle, which may have made the transmutation of species seem more plausible to their adherents in the Scotland of the 1820s.

## NOTES

1. Brown, 'Smellie, William', p.2.
2. Eddy, *The Language of Mineralogy*, p.165.
3. Ibid., p.32.
4. Ibid., p.130.
5. Ibid., p.44.
6. For a classic account of the history of this idea, see Lovejoy, *Great Chain of Being*.
7. Bonnet, *Contemplation de la Nature*, p.27.
8. Pope, *An Essay on Man*, p.9.
9. Lovejoy, *Great Chain of Being*, p.242.
10. Smellie, *Philosophy of Natural History*, vol.1, p.520.
11. Ibid., p.55.
12. Brown, 'Smellie, William', p.3.
13. The Latin expression that ends this quote can be translated as 'Nature never makes jumps'. Walker, Notes on natural history lectures (1791), f.37.
14. Smellie, *Philosophy of Natural History*, p.55.
15. Buffon, *Natural History, General and Particular*, vol.8, pp.62–3.
16. Smellie, *The Philosophy of Natural History*, p.523.
17. Walker, Notes on natural history lectures (1782), f.7 verso.
18. Walker, Notes on natural history lectures (1790), vol.4, f.162.
19. Walker, Notes on natural history lectures (1797), vol.9, f.135.
20. Ibid., f.164.
21. Buffon, *Natural History, General and Particular*, vol.4, pp.70–1.
22. Monro, Notes on lectures on anatomy (1774/5), f.197.
23. Monro, Notes on lectures on comparative anatomy (1786), f.10.
24. Translation by H. F. Roberts. Quoted from Glass, 'Heredity and variation', in Glass, Temkin and Straus, *Forerunners of Darwin*, p.149.
25. Walker, Lecture notes (1790), vol.4, f.163.
26. Walker, Lecture notes (1797), vol.9, f.140.
27. Monro, Lecture notes (1774/5), f.197.
28. Walker, Lecture notes (1790), vol.4, ff.168–9.
29. Glass, 'Heredity and variation', in Glass et al., *Forerunners of Darwin*, p.158.
30. Walker, Lecture notes (1791), f.19.
31. Ibid., f.19.
32. Walker, Lecture notes (1790), vol.4, ff.167–8.
33. Glass, 'Maupertuis', in Glass et al., *Forerunners of Darwin*, pp.70–1.
34. Walker, Lecture notes (1797), vol.3, f.135–6; Maupertuis, 'Essai de la Cosmologie', in Maupertuis, *Œuvres de Monsieur de Maupertuis*, pp.35–6.
35. Walker, Lecture notes (1790), vol.4, f.159.
36. Walker, Lecture notes (1782), f.7 verso.
37. Walker, Lecture notes (1791), f.45.

38. Walker, Lecture notes (1790), vol.4, f.161.
39. Ibid., f.160.
40. Harvey's dictum can be translated as 'every animal from an egg'.
41. Ibid., ff.161–2.
42. Walker, Lecture notes (1797), vol.9, ff.128–9.
43. Jameson, *Biographical Memoir of the Late Professor Jameson*, p.6.
44. Ibid., p.6.
45. Ibid., p.8.
46. Jameson, quoted in Sweet, 'Robert Jameson's Irish journal, 1797', p.110.
47. Sweet and Waterston, 'Robert Jameson's approach to the Wernerian theory of the earth, 1796', pp.81–95.
48. Eyles, 'Robert Jameson and the Royal Scottish Museum', p.157.
49. Dean, 'Jameson, Robert'.
50. Darwin to Hooker, 29 [May 1854], Darwin Correspondence Database.
51. Darwin, *Autobiographies*, pp.25–6.
52. Forbes, 'Professor Forbes's inaugural lecture'.
53. [Wakley], 'Biographical sketch of Robert Edmond Grant', p.689.
54. Quoted in Eyles, 'Robert Jameson and the Royal Scottish Museum', p.159.
55. Scottish Universities Commission, *Report Relative to the University of Edinburgh*, p.46.
56. Ibid., p.46.
57. Ibid., p.47.
58. Jameson, 'Syllabus of lectures on natural history' (1826), in Scottish Universities Commission, 'Appendix to No. XIII', *Returns, Papers, and Examination*, p.14.
59. Ibid.
60. Knox, 'An account of the *Foramen centrale* of the retina'; Knox, 'Observations on the duck-billed animal of New South Wales'; Knox, 'Additional observations relative to the *foramen centrale*'; and Knox, 'Inquiry into the origin and characteristic differences of the native races'.
61. Scottish Universities Commission, *Minutes of Evidence Taken before the Commissioners*, p.100.
62. Scottish Universities Commission, *Report Relative to the University of Edinburgh*, p.91.
63. Quoted in Duns, 'Memoir', in Fleming, *Lithology of Edinburgh*, p.xl. The reference to the journal that ends this quote is to Thomas Thomson's *Annals of Philosophy*.
64. [Cheek], 'On the present state of science abroad: No.1 Scientific coteries of Paris', p.118.
65. Cheek, *Answer to Certain Statements*, pp.3–4.
66. [Cheek], 'Miscellaneous intelligence: Edinburgh University', p.77.
67. Scottish Universities Commission, *Evidence, Oral and Documentary. Volume 1. University of Edinburgh*, pp.629–31.

68. Neill, *Supplement to an Address to the Wernerian Natural History Society*, p.4.
69. Ballingall, 'The life of Dr Barclay', in Barclay, *Introductory Lectures*, p.v.
70. Barclay, *New Anatomical Nomenclature*, p.1.
71. Barclay, *Life and Organization*, pp.236–7.
72. Spurzheim to Combe, 31 December 1831, f.63.
73. Barclay, *Life and Organization*.
74. Ibid., p.x.
75. Richards, 'The "moral anatomy" of Robert Knox', p.377.
76. Knox, 'Notice respecting the presence of a rudimentary spur', pp.130–2.
77. Knox, 'Inquiry into the origin and characteristic differences of the native races', p.210.
78. Knox, *Races of Men: A Fragment*, p.90.
79. Grant, *Dissertatio Physiologica Inauguralis*, p.8.
80. Grant, *Tabular View*, p.v.
81. [Wakley], 'Biographical sketch of Robert Edmond Grant', p.689.
82. Ibid., p.690.
83. Ibid., p.691.
84. Fleming, *History of British Animals*, p.524.
85. Grant, Essays on medical subjects, ff.88–179.
86. [Wakley], 'Biographical sketch of Robert Edmond Grant', p.692.
87. Desmond and Parker, 'The bibliography of Robert Edmond Grant', p. 204.
88. Desmond, *Politics of Evolution*, p.71.
89. Hutton Balfour, *Biographical Sketch of the Late John Coldstream*, p.13.
90. Darwin, 'Recollections of the development of my mind and character', p.23.
91. See, for example, Desmond and Moore, *Darwin*, p.41.
92. Darwin, 'Recollections of the development of my mind and character', p.24.
93. Desmond and Moore, *Darwin*, p.31.
94. Scottish Universities Commission, *Minutes of Evidence Taken before the Commissioners*, p.92.
95. Ibid., p.92.
96. Baird, *Memoir of the Late Rev. John Baird*, p.63.
97. Ibid., p.63; Anon, 'Local scientific societies', p.38.
98. Anon, Abstract of the Proceedings of the Plinian Society, p.33.
99. Anon, 'Natural history in Scotland', pp.291–2.
100. Minutes of the Plinian Society, vol.1, f.25 verso.
101. Anon, 'Natural history in Scotland', pp.291–2.
102. Minutes of the Plinian Society, vol.2, f.98 recto.
103. Ibid., f.99 verso.
104. Ibid., f.125 verso.
105. See, for example, Gruber, *Darwin on Man*, p.39; Desmond and Moore, *Darwin*, p.38; and Brown, *Charles Darwin: Voyaging*, p.77.
106. Scull, 'Browne, William Alexander Francis'.
107. Anon, 'Dinner by the Phrenological Society to Dr Spurzheim', p.142.

108. Minutes of the Plinian Society, 1826–1841, vol.1, f.57.
109. Ibid., f.11 recto.
110. Ibid., f.99 verso.
111. Ibid., f.71 recto.
112. Gray, *History of the Royal Medical Society*, p.73.
113. Ibid., p.73.
114. Ibid., p.74.
115. Stewart, 'On the different theories of the earth', f.507.
116. Ibid., p.508. Stewart is referring here to the race of philosophers who inhabit the flying island of Laputa in Part 3 of Jonathan Swift's *Gulliver's Travels*. Swift intended Laputa as a sature on the Royal Society, which he considered as a forum for idle and fruitless speculation.
117. Anon, 'Some account of the Wernerian Natural History Society of Edinburgh', p.233.
118. Minutes of the Wernerian Society, vol.1, p.232.
119. The change in name was necessitated by a change in publisher after the bankruptcy of Archibald Constable in 1826 and the transfer of the journal to his rival Adam Black.
120. The story of the split between Brewster on one side and Jameson and the journal's publisher, Archibald Constable, on the other, has been comprehensively covered in an enlightening article by Jonathan R. Topham, 'The scientific, the literary, and the popular', and also by W. H. Brock in 'Brewster as a scientific journalist'.
121. It was, for example, regularly referred to in this way in the pages of the *Magazine of Natural History* in the 1830s, and by Robert Chambers in both his *Vestiges of the Natural History of Creation*, p.172, and *Ancient Sea-margins*, p.286.
122. [Brewster], 'Advertisement'.
123. Ibid., p.viii.
124. Title page, *Edinburgh Journal of Natural and Geographical Science* 3 (1831), p.1.
125. Withers, 'Towards a history of geography', p.63.
126. Neill, *Supplement to an Address to the Wernerian Natural History Society*, p.16.
127. Neill, *Supplement to an Address to the Wernerian Natural History Society*, pp.15–16.
128. For an excellent account of this controversy, see Robertson, *Patrick Neil*, pp.46–7.
129. Ainsworth, 'A descent into Eldon Hole, in the Peak of Derbyshire', p.260. I am indebted to Julian F. Derry for bringing this fascinating article to my attention.

# 4

## Geology and Evolution

In November 1850, when John Fleming gave his inaugural lecture on natural science at the newly established New College of the Free Church in Edinburgh, where he was to be the first professor of natural history, he used the occasion to settle an old score. Included in his address was an attack on the University of Edinburgh's aged professor of natural history, Robert Jameson. Fleming made very clear that he associated Jameson, and the Wernerian geology he had long espoused, with the evolutionary speculations that he had come to despise. In his lecture he said:

> Subsequent to the rise of this Scottish geology of Hutton, the German geology of Werner was introduced, and for a while appeared to triumph. This system, equally indifferent to the truths of palaeontology, and outraging all philosophy by the extravagance of its assumptions, paved the way for those reveries of progressive development with which of late years we have been inundated.[1]

Jameson had been the leading champion of Wernerian geology in the English-speaking world, so to associate Werner's theories with evolutionary speculation was, in Fleming's opinion, an effective way of besmirching the old professor's reputation. Fleming also made a point of emphasising the foreignness of Werner and his system in contrast to the Scottish geology of Hutton. By the 'reveries of progressive development', Fleming alluded to Robert Chambers' anonymously published evolutionary magnum opus, *Vestiges of the Natural History of Creation*, which had caused enormous controversy on its publication in 1844. Although it had been published some years before Fleming made his address, Chambers' book was still extremely controversial in 1850 when Fleming gave his lecture. It emerged later in his lecture that the immediate cause of Fleming's ire had been that Jameson had given broadly favourable reviews to both *Vestiges* and its Chambers' follow-up volume *Explanations: A Sequel to 'Vestiges of the Natural History of Creation'* (1845) in the pages of the *Edinburgh New Philosophical Journal*.

Jameson had this to say in his review of *Vestiges*: 'Although we do not agree with the ingenious author of this interesting volume in several of his speculations, yet we can safely recommend it to the attention of our readers'.[2] The following year he reviewed *Explanations*, noting that 'These explanations sufficiently prove that the author has met with great effect the arguments of its distinguished opponents'.[3] While perhaps not meriting the term 'laudatory', Jameson's reviews at the very least indicate that he maintained an open mind towards the 'development hypothesis' that Chambers expounded in *Vestiges* and that Fleming referred to dismissively as 'reveries of progressive development'. It is not surprising that Jameson was less than enthusiastic in his praise of *Vestiges*, assuming he had not changed his views on evolution substantially since the late 1820s. Chambers' theory relied on an innate tendency to progressive development instilled into living things by the deity at the creation. As we will see below, this is a very different model of evolution from the one espoused in the circle of natural historians around Jameson. That Jameson considered *Explanations* to have effectively answered the arguments of the critics of *Vestiges* must have been particularly galling to Fleming, as these critics included Fleming himself, as well as and his friend and fellow evangelical David Brewster, who wrote a blistering review of *Vestiges* for the *North British Review*.[4] Fleming roundly condemned Chambers' book in the strongest possible terms later in his inaugural lecture, and it seems that in his mind there was no doubt regarding the links between the scandalous theory of universal progress outlined in that work and the developmental vision of the history of the earth advanced by the Wernerians earlier in the century. In this chapter I will explore the extent to which Fleming was right to make a strong association between the Wernerian theories that dominated geology in Scotland in the early decades of the nineteenth century and evolutionary speculation. In particular I will be investigating the role played by Robert Jameson in the development and propagation of progressive, or even evolutionary, models of the history of life on earth.

## THE WERNERIAN MODEL OF EARTH HISTORY

In a footnote to the introduction to his textbook of Wernerian Geology, *A System of Mineralogy* (1804), Robert Jameson laid out what he saw as the main problems of natural history that remained to be solved as the nineteenth century began. For Jameson, the most important questions included: 'Were all animals and plants originally created as we at present find them, or have they by degrees assumed the specific forms they now possess? Are certain species become extinct? In what order and whither have they migrated? What change has climate produced?'[5] Right at the very beginning of his professorial career, Jameson was already raising important questions regarding the history of life on earth. First among these questions was that of the origin and evolution of species. Like most of Jameson's works, his *System of Mineralogy* was deeply rooted in the theories of Abraham Gottlob Werner. It was against a background of the Wernerian theory of the earth that these questions emerged to trouble Jameson. It is therefore to Werner's theories, as understood and developed by Jameson, that we must first turn in order to establish what answers, if any, Jameson found for the questions he had posed in 1804 in the decades that followed.

Werner had taught that the oldest rocks in the earth's crust were deposited in a universal ocean, very different in chemical composition from the present oceans of the world. The spherical shape of the earth was taken by Wernerians as evidence for its original fluidity.[6] The subsequent geological history of the earth was essentially the story of the gradual retreat of this ocean and the subsequent alterations in the physical conditions that accompanied this process. This in itself was not a new idea. Benoît de Maillet had put forward a model of earth history based on a gradually retreating ocean in the early eighteenth century. As a consequence of his theory he had also proposed that all terrestrial life had descended from marine forms. Werner's theory was, however, considerably more elaborate and sophisticated than de Maillet's speculations. Werner and his many disciples believed that the oldest, or 'primitive', rocks had been deposited by chemical precipitation from the primordial ocean. These rocks were crystalline, silicate rocks, such as granite and gneiss. The Wernerians concluded that they must have been deposited first, as they were aware that silica was soluble only in hot, chemically basic liquids, of the kind they imagined constituted the primordial ocean.[7] Subsequent rocks of the 'transition' and succeeding 'floetz' periods were deposited as the waters receded from the emerging continents. As dry land emerged, rocks were increasingly deposited by mechanical deposition as material was eroded from the land and washed into the oceans. As time went by these mechanically deposited rocks, such as sandstone and shale, came to predominate, although chemically precipitated rocks, such as limestone, continued to be formed. The most recent 'alluvial' deposits were almost entirely mechanically formed, the youngest of them still being largely unconsolidated. As would be expected from their deposition from the waters of a gradually receding universal ocean, the oldest rocks formed the cores of high mountain ranges, while the alluvial deposits were to be found in low-lying areas. What had happened to all the water was a major problem for the theory, and was a question constantly raised by its opponents. A variety of different hypotheses existed among Wernerian geologists as to the cause of the recession of the ocean. The one favoured by Robert Jameson was that the water had been lost to space over the millennia.[8]

Many Wernerians, Jameson included, believed that in addition to declining sea levels there had also been a continuous diminution of the temperature of the earth over geological time. This doctrine is often associated with Buffon, who saw the cooling of the earth from an original molten state as the primary motor for change in the history of the earth. The section of the natural history syllabus that covered botany for the 1826 session included 'Deductions illustrative of Gradual Change in the Heat of the Earth, and of Alteration in Climate, as disclosed by the facts in the Physical and Geographical Distribution of Fossil and Living Plants'.[9] We can learn from a set of student's notes from 1830 what Jameson had to say in this lecture on the direction and effects of climatic change. He suggested that, in the geological past, 'the climate was very different from what it is at present and that at the time Britain was calculated to produce plants and animals requiring a much more considerable temperature then the Island possesses at present'.[10] In a fragment of manuscript found among Jameson's papers we also find the following note in Jameson's hand that suggests he saw a diminution in temperature over geological time as not only likely but compatible with a broadly Wernerian picture of earth history:

> 7 [owing?] to the diminution of temperature the expanded state of water becoming less the quantity of the atmospheric vapor & height of water diminishes
>
> 8 the Transition rocks the next down from them conglomerate, charcoal & their organic remains intimate the existence of mechanical action & of such a temp as to allow of the growth of organic bodies all of which appear to have been marine[.][11]

Werner's theories were enormously influential across Europe. In the English-speaking world Jameson was their foremost champion and promoter. Rachel Laudan has demonstrated that the Werner's neptunian theory of the earth 'dominated geology until the late 1820s'.[12] As can be seen from the surviving notes from his lectures, it was essentially the Wernerian model of earth history that Jameson taught his students from 1804 until at least the mid-1830s.[13] The belief that the geological record could be explained by progressive, directional change in the conditions on the surface of the earth became so dominant in the early nineteenth century that Martin Rudwick has referred to theories of the type espoused by Werner as the 'standard model' for the period. The main opponents of the Wernerians were the followers of James Hutton. The city of Edinburgh was the centre of propagation in Britain for both of these theories, so it is hardly surprising that their advocates should come into conflict there. I will be turning to the dispute between the followers of these two rival theories of the earth in the next section.

## WERNERIANS AND HUTTONIANS IN EDINBURGH

The situation in geology at the beginning of the nineteenth century has been accurately summarised by Martin Rudwick, who has noted that the theories prevalent at the time could be classified into two categories: 'those that postulated an earth in steady state or cyclic equilibrium and those who saw the earth's temporal development in directional terms'.[14] Werner and his followers belonged to the latter camp, while the theory developed by James Hutton (1726–97) represented an extreme example of the former. (As we will see below, Jean-Baptiste Lamarck also advocated a steady-state theory of the earth, although one with fundamental differences from Hutton's.) While Werner's theory interpreted the geological record as showing a clear pattern of progressive change over time, Hutton's theory was radically ahistorical, centred on a uniformitarian model of the history of the earth, with, as he famously put it, 'no vestige of a beginning, – no prospect of an end'.[15] For Hutton the earth's history was an endlessly repeated cycle of uplift and erosion. The inner fires at the centre of the earth raised up strata through volcanic action to become continents, only for them to be eroded and carried down into the depths of the oceans again by gravity and the action of water. These sediments would then be consolidated into rock through the action of the earth's internal heat and uplifted again to start a new cycle. This model of the history of the earth fundamentally determined the way Huttonians interpreted the geological record. Hutton believed in an essentially Newtonian vision of the earth. Just as the planets eternally orbited around the sun, Hutton's 'world machine' was also in perpetual motion, powered by the opposing principles of the attractive force of gravity and the expansive power of heat. But for religious considerations, there would have been no reason to assume that this dynamic but unchanging natural order was not eternal.

For Wernerians different types of rocks related to different phases in the history of the earth. Crystalline rocks such as granite represented the oldest phase of earth history, while unconsolidated alluvial deposits were the most recent. But for Huttonians, any type of rock could be of any age; the crucial differences between rock types related to the amount of heat and pressure they had been exposed to and had nothing to do with the specific epoch in which they were formed. James Hall (1761–1832), one of

Hutton's most fervent Scottish supporters, reported a series of experiments to the Royal Society of Edinburgh in June 1805 that claimed to demonstrate that the action of heat and pressure could transform sedimentary rocks in just the way postulated by Hutton.[16] His results were, of course, controverted by the Wernerians. Thus Hutton's timeless order in perpetual motion confronted Werner's earth with its definite beginning and a directional trajectory towards its end. Edinburgh was the principal backdrop against which the conflict between the two visions of earth history was to play out.

The most important of Hutton's followers in Edinburgh were John Playfair and Thomas Hope. Playfair had written *Illustrations of the Huttonian Theory of the Earth* (1802), which presented Huttonian geology in a more accessible form than Hutton's own rather daunting and difficult 'Theory of the Earth' (1788). Hope was Jameson's main Huttonian rival within the University. When the Scottish Universities Commission quizzed Jameson about their rivalry in 1826, he made light of their disagreement, saying: 'It would be a misfortune if we all had the same way of thinking; Dr. Hope is decidedly opposed to me, and I am opposed to Dr. Hope, and between us we make the subject interesting'.[17] Although Jameson could make light of the dispute between Wernerians and Huttonians in 1826 before the Scottish Universities Commission, hostilities between the two parties at times stopped little short of physical violence. The most extraordinary and notorious incident in the dispute between the two schools of thought took place in January 1812 at Edinburgh's Theatre Royal.[18] George Mackenzie (1780–1848), an ardent Huttonian, had recently returned from a trip to Iceland, where he had gathered evidence in favour of the Huttonian theory of the earth. During his trip he also took an interest in the folklore of the country, and on his return he wrote a play based on a traditional tale entitled *Helga and Her Lovers*. Sadly for the playwright, the opening night was marred by a group of noisy and aggressive Wernerian geologists, intent on disrupting the play, although the piece itself seems to have had no discernible geological content. They succeeded all too well, and the play was forced to close after the first night. That such an apparently arcane scientific dispute could cause uproar in a public theatre goes to show what strong feelings geology could arouse in Edinburgh in the early decades of the nineteenth century. Given the notoriously high spirits of Edinburgh students in the period, it must be strongly suspected that some of Jameson's young disciples from his natural history class were involved.

Both the Wernerian and Huttonian theories found their partisans among the students of the University. The rivalry between the two subjects was made the subject of a number of essays read by student members of both the Royal Medical Society and the Royal Physical Society. There is some evidence to suggest that Jameson had already made converts among the student body to Wernerian geology and its connection with a progressive vision of the history of life in his first few years as professor of natural history. In the dissertation book of the Royal Medical Society for 1806/7 can be found an essay by a student of Jameson named J. Ogilvy entitled 'On the Huttonian and Neptunian theories of the earth'. In this dissertation Ogilvy praised the professor's 'masterly statement of the Wernerian theory'.[19] He made his contempt for the Huttonians explicit in his essay, in which he admitted to '[h]olding the Plutonic [Huttonian] theory so cheap, and conceiving its founder to be altogether destitute of that knowledge which should proceed Geology' that he considered it a waste if his time to discuss its details, 'being ready to prove in debate, that almost every fact advanced by the Huttonian to support his system, is either founded on inaccurate observation, or admits of easy solution by the water of the Neptunian'.[20] H. Gilby provided a Huttonian riposte in the dissertation book for 1813/14, in which he criticised Jameson for failing to provide a satisfactory explanation for the retreat of the oceans.[21] Of the four essays given to the Royal Physical Society between 1800 and 1823 dealing with this great geological controversy, three in the end concluded that all theorising in geology was in fact a waste of time, whether Huttonian or Wernerian. John C. McDowell in the volume for 1806/7, for example, concluded that: 'With regard to Geology; therefore mankind are still groping in the dark; – a darkness which Theories, & particularly the igneous tend only to increase'.[22] The author of the one fully committed essay, J. William Watson, on the other hand was a staunch Wernerian. He wrote:

> Dr Hutton then in proposing his theory, seems to have overleapt the boundaries of his own assertions and lost sight of his own principles which suppose, that all the strata of which the globe is composed have consolidated by means of heat: – that the exhibition of the common ordinary phenomena of heat is not to be looked upon in the grand process of nature I conceive only the facts I have now produced are sufficient to shew, and I hope the day is not far distant when the theory of the noble and illustrious Werner will be found, by every unprejudiced Geologist, as the only one founded on true and rational principles.[23]

Although Hutton's theory was to play an important role in the later history of geology, principally in the form in which it was recast by Charles Lyell (1797–1875), in 1820s Edinburgh it was the Wernerian model that seems to have triumphed. It owed its success in large part to the indefatigable energy of its most prominent apostle, Edinburgh's professor of natural history, in promoting it among his students and fellow natural historians. It must also have benefited from the growing realisation that the fossil record showed clear evidence of directional change, which was hard to explain in terms of the endless cycles of Huttonian geology. For Hutton, life on earth, including the human race, also constituted an effectively eternal, unchanging natural order.[24] As the leading Huttonian, John Playfair, was to write, consciously echoing Hutton's famous adage quoted above, '[i]n the continuation of the different species of animals and vegetables that inhabit the earth, we discern neither a beginning nor an end'.[25] This model increasingly jarred with the evidence for directional change being uncovered in the fossil record, a tendency that made much more sense when viewed through the lens of Wernerian geology.

## THE STORY OF LIFE AS A TALE OF PROGRESSIVE DEVELOPMENT

By the early decades of the nineteenth century there was a growing consensus among geologists that there was a trend towards greater variety and complexity in the fossil record on passing from older to younger strata. They also noticed that the major groups of living things seemed to make their appearance in a defined order. First to appear were simple invertebrate animals, followed by fish, then terrestrial animals and birds and finally man, whose remains had not yet been found in fossil form. The majority concluded from this that the history of life was indeed progressive and by the early nineteenth century this had become the consensus among most geologists.[26] Robert Jameson was no exception; in a set of notes taken by a student in Jameson's lectures we find a clear exposition of this succession of fossil forms. These notes are undated, but did not originate earlier than 1826, as they contain a reference to a paper published that year. In them we find the following outline of the fossil record:

> In the oldest strata [of the Transition rocks] we find the lowest species of vegetables & animals, as marine plants and zoophytes, which were therefore first called into existence ... Floetz rocks are less crystallized than transition rock, but contain a greater variety of organic remains. Indeed there appears to be a regular & consistent distribution of organic beings through the rocks of this class from the very low species of the earliest strata to the more perfect animals of the newest strata, immediately adjoining the alluvial formation.[27]

Jameson, like other Wernerian geologists, related the progressive nature of the history of life to the directional changes in the physical conditions on the surface of the earth that were integral to their theories. Gradual changes in the chemical composition of the oceans and the progressive reduction in the temperature of the earth made it increasingly hospitable to life and allowed new forms to appear and flourish. This led directly to the trend towards increasing variety and complexity so apparent in the fossil record. In his *Elements of Geognosy* (1808) Jameson wrote:

> As the water diminished, it appears to have become gradually more fitted for the support of animals and vegetables, as we find them increasing in number, variety and perfection, and approaching more to the nature of those in the present seas, the lower the level of the outgoings of the strata, or, what is the same thing, the lower the level of the water. The same gradual increase of organic beings appears to have taken place on the dry land.[28]

Thus contemporary interpretations of the fossil record came to the defence of the Wernerian model of the history of the earth. Not only the progressively changing nature of the rocks themselves, but the gradual appearance of increasingly complex and varied living things in the fossil record bore witness to a physical environment subject to continuous, directional change of just the kind proposed by the Wernerians.

A belief in the existence of such a connection between geological and biological development was instilled by Jameson in his student disciples. J. Ogilvy, the Wernerian medical student and member of the Royal Medical Society mentioned above, compared the Wernerian and Huttonian theories in the dissertation he read to the Society:

> Another general observation of the same philosopher [Werner] beautifully confirming his opinion, – is the constantly increasing frequency of the relicts of the animal and vegetable kingdoms, as we descend from the Transition to the rocks of most recent formation; and, at the same time, he sagaciously remarked, that, in making this descent, these vestiges point out individuals of these kingdoms with which we become the more familiar as we approach the most modern formations.[29]

Some more established natural historians also shared Jameson's views. We have seen that in 1850 John Fleming proved himself a harsh critic of both Werner and Jameson, as well as a resolute opponent of evolutionary speculations. However, Fleming's earlier writings tell a very different story. His published work up until the early 1820s shows that at that time Fleming had shared Jameson's Wernerian opinions; he had even been a founding member of the Wernerian Natural History Society when it was established by Jameson in 1808. In his *Philosophy of Zoology*, published in 1822, he wrote:

> From the period, therefore, at which petrifactions appear in the oldest rocks, to the newest formed strata, the remains of the more perfect animals increase in number and variety; and it is equally certain, that the newest formed petrifactions bear a nearer resemblance to the existing races, than those which occur in the ancient strata.[30]

Fleming had believed, as did Jameson, that these progressive changes found in fossil faunas were related to the evolution of the terrestrial environment. Like Jameson, he had believed that the area of dry land had increased over time at the expense of a primordial universal ocean, although he suggested that this had occurred through the filling in of lakes and seas by the products of erosion rather than by a net loss of water. Although he may have differed from Jameson about the details of the mechanism driving environmental change, he expressed its consequences for life on earth in a similar manner:

> A variety of changes have taken place in succession, giving to the earth its present character, and fitting it for the residence of its present inhabitants. And if the same system of change continues to operate, (and it must do while gravitation prevails,) the earth may become an unfit dwelling for the present tribes, and revolutions may take place, as extensive as those which living beings have already experienced.[31]

Jameson's *Edinburgh New Philosophical Journal* provided a forum for Wernerian geologists to exchange ideas about the history of the earth and interpretations of the fossil record. Articles published there often touched on the relationship between the progressive changes in the physical conditions on the surface of the earth and the history of life, as made manifest in the fossil record. The author of one important article published in 1826 under the title 'Geological observations' was the Wernerian geologist and former student of Jameson, Ami Boué. In his paper, Boué observed that 'the farther we penetrate into the crust of the earth, the more simplicity do we observe in the vegetable and animal productions'. He then went on to suggest that this progressive change was due to a greater equality of temperature across the globe in the past, concluding that as 'the zones and climates gradually became established, the vegetables and animals became diversified'.[32]

The following year Jens Esmark (1762–1839), the Danish-Norwegian Wernerian mineralogist, suggested in a paper in Jameson's journal that the earth might have been devoid of life for several thousand years after the creation, and that 'organisation did not begin till this long period was completed, which the earth required to the full development of its own constitution; that, after it began, it proceeded by successive steps from the less to the more perfect formations, ending with man as the head of the whole'.[33] Esmark presented a model in which the first appearance of life was made possible by changes in the physical conditions on the earth, without clarifying whether these new conditions actually brought life into being. This progressive change then promoted the increase in the complexity of living things that had been observed in the fossil record. Jameson was clearly favourably impressed by Esmark's article; we know from a set of notes taken by one of Jameson's students that survives in Edinburgh University Library that he discussed this article approvingly and at some length, describing it as a 'very ingenious paper'.[34]

Another paper that presented a very similar model of the relationship between the history of the earth and the history of life appeared in Jameson's journal in 1830. In this case it was published anonymously. The picture it paints of gradual, progressive development of life is more or less the same as that depicted in the earlier papers by Boué and Esmark. The anonymous author notes:

> It is, notwithstanding, always of much importance to be able to look into the facts already established, and to observe that the gradual development of organic bodies in the animal and vegetable kingdom has followed precisely the same progress. While the simplest organized kinds of both kingdoms first appear, we also find repeated throughout the same gradations, as regards the gradual appearance and increase of the most perfectly organized beings in the strata of the earth's crust.[35]

Jameson continued to publish articles articulating a progressive model of the history of life into the 1830s. In 1835 the *Edinburgh New Philosophical Journal* published a paper by the French-Swiss botanist Alphonse de Candolle (1806–93) on the fossil history of plants that shared this progressive view of the development of life. In it de Candolle stated:

> With these results before us, we can recognise with M. [Alexandre] Brongniart, that the greater number of organs, and those the most distinct, have succeeded to the less perfect ones; in other words, that the vegetable kingdom appears to have been gradually becoming more perfect. This law of gradual development would hence appear to exist in the vegetable, as it has been supposed to exist in the animal kingdom.[36]

De Candolle here goes as far as to posit that the exitence of a 'law of gradual development' explains the evidencing of increasing complexity in the fossil record, without going so far as to propose a mechanism to explain how this law operates.

Another late contribution on the theme of the progressive history of life on earth appeared in volume 7 of the *Memoirs of the Wernerian Natural History Society*, which covers the period 1831–37. Here we find a lengthy article by Robert James Hay Cunningham (1815–42) presenting an unambiguously progressive picture of the fossil record. This article, entitled 'On the geology of the Lothians', won the prize offered by the Society in 1836 for the 'best geological report of the Lothians'. In his paper Cunningham stated:

> If geologists have, in the course of their investigations, come to any certainty concerning the ancient state of our globe, there is certainly no one doctrine supported by a greater number of facts, than that of progressive development. Many remains have been adduced as belonging to beings, which held a place in the zoological scale, higher than was consistent with this theory. With one exception, however, all these remains have been found, on more accurate and better conducted examination, to be, instead of dissentient facts beautiful proofs of its truth.[37]

Cunningham was in no doubt that this one controversial exception, the famous Stonesfield mammal found in sediments that, on the basis of the other fossils they contained, were considered too old to contain mammalian remains, would sooner or later be satisfactorily explained. While no mechanism is proposed here for the 'progressive development' evident in the fossil record, this paper is clearly open to an evolutionary interpretation. It was presumably on these grounds that John Fleming was later to criticise the progressivist aspect of this paper in his *Lithology of Edinburgh* (1859).[38]

Jameson's publishing endeavours, and in particular the *Edinburgh New Philosophical Journal*, allowed him to provide a platform for Wernerian geologists to speculate about the relationship between directional change in the physical environment and the progressive story of life revealed by the fossil record. It is evident from them that the link between a directional history of the earth and a directional history of life was clearly a commonplace of the geological circles around Jameson. While all the articles examined in this section suggested that the appearance of new forms of life was associated with changes in the physical conditions, none of them go so far as to speculate on the relationship between the older, more primitive forms of life and their more advanced successors. They all stopped short of any suggestion that the pattern observed in the fossil record was a consequence of the transmutation of species. It is therefore impossible to say if any of these authors considered that an evolutionary explanation of the trend towards increasing complexity and diversity was evident from the fossil record. The sources I will be considering in the next section show that at least some of the natural historians in the circle around Robert Jameson had indeed come to look on the progressive history of life as the consequence of evolutionary development.

## WERNERIAN GEOLOGY AND TRANSFORMISM

In his *Philosophy of Zoology* (1822) John Fleming gave surprisingly serious consideration to an evolutionary interpretation of the story told by the fossils. He wrote that it had been suggested by some natural historians that 'the present races of animals and vegetables, are the descendants of those whose remains have been preserved in the rocks'.[39] He then cited the ease with which man had modified domestic animals and plants as strong evidence in favour of this conjecture. However, he went on to reject the transformist model on the basis of two arguments. First, he claimed that there were limitations on the degree to which living things could be modified by their physical conditions of life. Second, he argued that the existence of the intermediate forms that would give the theory credibility had yet to be established. Unlike the figures discussed in the last section of this chapter, Fleming did openly consider a transformist explanation for the history of life, even though he explicitly stated his scepticism regarding the transmutation of species. Nonetheless, it is clear that, at least in the early 1820s, he was still prepared to give its advocates a hearing.

As well as considering an evolutionary explanation for the progressive trend evident in the fossil record, Fleming also raised another intriguing possibility. In his *Philosophy of Zoology* he suggested that:

> If the seeds of some plants, and the eggs of certain animals, be so minute as to be excluded with difficulty from any place to which air and water have access, and if they are capable of retaining, for an indefinite length of time, the vital principle, when circumstances are not favourable to its evolution, the crust of the earth may be considered as a mere receptacle of germs, each of which is ready to expand into vegetable or animal forms, upon the occurrence of those conditions necessary to its growth.[40]

Fleming's explanation is reminiscent of Buffon's theory of 'organic molecules', which he believed were ubiquitous in the natural world, waiting to be organised into a living being by an appropriate 'internal mould'. Both Fleming and Buffon seem to have believed that there might exist active principles in the natural world that could perform the function of Buffon's 'internal mould' and facilitate the development of individuals of a previously unknown species from the ubiquitous 'germs' or 'organic molecules'. As the physical conditions became suitable for new forms of life to survive and thrive, these new species would therefore arise spontaneously. This theory was presented by Fleming as a possible rival, non-evolutionary explanation for the progressive appearance of new species in the fossil record. His discussion of this second non-evolutionary model provides a valuable warning that it is not safe to read transformist ideas into any expression of belief in a progressive fossil record without concrete evidence that the author intended this progression to be understood in terms of descent with modification. While paeans to progress in the natural world and the fossil record that were published in the *Edinburgh New Philosophical Journal* and elsewhere can, when viewed through post-Darwinian spectacles, be read as evolutionary tracts, an expression of belief in a progressive history of life cannot be taken as acceptance of the descent of one species from another in the absence of an explicit statement to that effect or strong external evidence. Fleming, for example, while still adhering to an essentially Wernerian model of the history of the earth and the progressive nature of the fossil record, could put forward a quite different hypothesis to explain this. While much of the published evidence discussed in the last section hinted strongly that an evolutionary interpretation had been in the mind of the author, and in some cases we know that the authors did openly express evolutionary opinions elsewhere, by not speculating as to the cause of the progressive nature of the fossil record these authors left open the possibility that some other mechanism was at work. It is for this reason that I have distinguished those sources that simply point out the progressive nature of the history of life from those that propose a specifically evolutionary explanation for this observed pattern of progressive development. A discussion of the significant body of material published in Edinburgh that did unambiguously express evolutionary ideas based on descent with modification will be the subject of the rest of this section.

I turn now to those crucial published sources containing explicit statements of their authors' belief that the species found in the fossil records were indeed the ancestors of more recent forms. The authors of these texts all work within an essentially Wernerian model of the history of the earth. One of the most important sources of evolutionary texts from the 1820s is Robert Jameson's *Edinburgh New Philosophical Journal*. The transformist contribution to Jameson's journal that has been most widely discussed by scholars is the anonymous 'Observations on the nature and importance of geology', published in 1826. As noted above, there has been some debate about the authorship of this article. Whether the author was Robert Grant, Robert Jameson, Ami Boué, or some other figure, James Secord has convincingly demonstrated that the author was almost certainly a Wernerian geologist. This is evident from the nature of his argument as well as the specifically Wernerian geological terminology he used. The most convincing attribution of the authorship of this article by James Secord, and now widely accepted by other scholars, is to Robert Jameson. Another plausible candidate suggested by Pietro Corsi is Ami Boué.[41] Most of the arguments in favour of Jameson would hold equally well for Boué, who had attended Jameson's classes when he was a medical student in Edinburgh between 1814 and 1817. Like Jameson, to whom he still referred in his autobiography many decades later as 'mon maître', he was a Wernerian geologist.⁴² The author focuses much of his argument on the crucial role played by changes in physical conditions in the history of life. The article notes the presence, in the rocks of colder parts of the globe, of fossils of species only found today in hot climates, indicating 'a great change in the temperature of their former situations'.⁴³ If this is so, the author maintains, it raises an important question about the effect that such changes have on living things. The changes that can be observed to have been wrought on domesticated plants and animals by modifying their conditions of life help to provide an answer:

> But are these forms as immutable as some distinguished naturalists maintain; or do not our domestic animals and our cultivated or artificial plants prove the contrary? If these, by change of situation, of climate, of nourishment, and by every other circumstance that operates upon them, can change their relations, it is probable that many fossil species to which no originals can be found, may not be extinct, but have gradually passed into others.⁴⁴

This passage makes clear that the author considers that directional change in physical conditions is the ultimate cause of the transmutation of species. Directional change in the surface of the globe, of the kind that is integral to the Wernerian model of the history of the earth, is therefore put at the centre of this evolutionary theory.

Even if there remain some doubts about the authorship of the anonymous 'Observations', there is significant evidence from other sources to suggest that Jameson was sympathetic to an evolutionary interpretation of the history of life. In his preface to the fifth edition of the English translation of Cuvier's *Theory of the Earth*, Jameson wrote that geology 'discloses to us the history of the first origin of organic beings, and traces their gradual development from the monade to man himself'.⁴⁵ In an appendix by Jameson entitled 'On the universal deluge' that was published in the same edition of Cuvier's book, he went on to add the following telling observation: 'like the formation of the rocks, we observe a succession of organic formations, the later always descending from the earlier, down to the present inhabitants of the earth, and to the last created being who was to have dominion over them.'⁴⁶ These passages would clearly seem to indicate that Jameson interpreted the succession of fossil forms found in the geological record in genealogical terms rather than as a series of progressive but separate creations. The relationship of fossil forms to living species was therefore that of ancestors to descendants.

There is also some evidence from Jameson's unpublished papers that he had been thinking about the transmutation of species from relatively early in his career. Among the papers held by Edinburgh University Library there is an incomplete manuscript in Jameson's handwriting containing a lengthy discussion of transformism. Unfortunately this is undated, but the paper it is written on bears an 1802 watermark, so it is likely to be early. The content also suggests an early date, as it harks back to eighteenth-century theories of transformism. The number and nature of the many amendments make it highly unlikely that he was simply copying from an English-language source, although the possibility cannot be entirely ruled out that he was translating from an unidentified foreign-language publication. The manuscript first discusses the process by which varieties become species over time:

> the different species kinds of Colewort which are now as varieties, may in process of time become fixed species – the different varieties of Dog as the Bull dog, Spaniel &c if kept separated from each other for centuries would form distinct species – that would not intermix with each other – like the Fox & Wolf which appear to have been formed from the Dog species in this manner[.][47]

According to this manuscript, the transmutation that produced new species had been brought about by environmental factors; these changes would then have been transmitted to offspring: 'Every peculiarity of climate, of nourishment, of generation, even many accidental mutilations, may give rise to these differences & thus form new varieties, which may pass into new[?] & fixed species after a long series of years.'[48] Jameson clearly considered that changes in the conditions of existence could lead to the origin of a new species, leaving open the possibility that the gradual, progressive change in physical conditions proposed by Werner could also drive change in the living world. This is not the only mechanism by which new species could come into being, as the 'number of species in the animal & vegetable kingdoms [illegible word] are formed not only by the transition of varieties into species, but also by the intermixture of different species'.[49] This latter evolutionary mechanism appears to be a reference to the theory proposed by Carl Linnaeus, who suggested that new species of plants could be generated through hybridisation. If this were indeed an original manuscript by Jameson, which seems highly likely, it would be very strong evidence indeed for his wholehearted acceptance of transformism at a relatively early date.

For reasons that I will discuss in the chapter that follows, it seems that Jameson may not have become familiar with the details of Lamarck's theory until the early 1820s. However, there is no doubt that he knew about older evolutionary theories long before then, and we know that as early as 1804 he had identified the origin of species as one of the key problems to be confronted in natural history in the nineteenth century. One context in which he would have heard about these theories would have been in the lectures of John Walker, his patron and predecessor as professor of natural history at Edinburgh. We know from surviving sets of student notes that Walker was dismissive of evolutionary theories, stating that: 'Transmutation of Species either in Plants or Animals, is a Vulgar Error'.[50] He did, however, clearly discuss evolutionary theories in some detail in his lectures. Most of the evidence we have from his lecture course relates to the theory of Linnaeus, which Jameson also referred to in the undated manuscript mentioned above. Jameson would certainly have been present when Walker discussed such eighteenth-century evolutionary theories in the course of his natural history classes. Later, armed with a superior knowledge of the fossil record than was available to Walker, he may have found reasons to question the opinion of his old professor. However, the evolutionary theories that Walker introduced only to dismiss as flights of fancy in his lectures may have spurred Jameson to begin to formulate his own answers to the problems posed by the appearance and disappearance of species that was so apparent from the evidence of contemporary geology.

In 1827, Jameson's journal published another article postulating an evolutionary interpretation of the fossil record entitled 'Of the changes which life has experienced on the globe'. This has received less attention from scholars than 'Observations on the nature and importance of geology', although it has been suggested by Adrian Desmond that it might have been by Grant.[51] While Grant is certainly a possible candidate as the author, it is not possible to confidently assign authorship. It is unlikely to be by Jameson, as the references to the important role of volcanism and 'the original igneous state of the earth' would be incompatible with his Wernerian views on the original aqueous state of the globe.[52] These suggest that the author had at least some sympathy with the Huttonian model of earth history. This would not rule out Boué, who himself admitted he was not as zealous a Wernerian as Jameson.[53] The article opens with a reference to the importance of fossils as evidence of 'the history and successive changes of the various races that existed before the present'.[54] The author then goes on to identify two types of causes at work in the natural world. The first and most important set of causes act gradually but inexorably: 'The differences which vegetables and animals exhibit at the present day, according to the various climates or situations in which they occur, have been gradually established under the predominating influence of a small number of natural causes, and constitute at length the order of distribution which life now presents at the surface of the earth.'[55] He then proceeds to expand on the nature of these causes:

> These gradual variations in the temperature, the lowering of the general level of the seas, the equally successive and gradual diminution of the energy of volcanic phenomena arising from the original igneous state of the earth, as well as the strength and power of atmospheric phenomena, and of the tides – such were the regular, general, and continued natural causes of the modifications which life has undergone . . .[56]

The author then calls the fossil record as witness to 'the successive and gradual change which we have pointed out'.[57] The second and less significant type of cause to which the author then turns consists of 'the irregular, and more or less violent and perturbing secondary causes of the partial vicissitudes experienced by animal and vegetable life'.[58] These secondary causes were the result of contingent factors, such as small-scale changes in the distribution of land and sea and the courses of rivers, and local climatic changes. The primary cause of transmutation was therefore gradual, directional change in external physical conditions. In this, the anonymous author seems once again to have been influenced by the Wernerian model of earth history. In the final section of the article the author expressed his overwhelming confidence in the correctness of his theory and appealed to its compatibility with natural law as confirmation: 'Our theory, which is founded on all the facts that have been established, cannot but prevail over the systems hitherto established, for it is in harmony with the natural laws of order and permanency which rule the universe.'[59]

Robert Grant was the only author to publish explicitly transformist articles in the *Edinburgh New Philosophical Journal* under his own name in the 1820s. Grant published sixteen papers in total in this journal and its predecessor the *Edinburgh Philosophical Journal* between 1825 and 1827, although the majority of them did not deal with transformist themes. These papers mostly discussed aspects of the biology of the invertebrates he had collected from the Firth of Forth, sometimes in the company of Charles Darwin. Two of these articles deal explicitly with the evolutionary history of the animals he studied. It is in a paper published in 1826, 'On the structure and nature of the Spongilla friabilis', that we find the first statement in print of Grant's transformist views in a discussion of the family relationship between marine and freshwater sponges:

> From this greater simplicity of structure and internal texture, we are forced to consider it as more ancient than marine sponges, and most probably their original parent; and, as its descendants have greatly improved their organization, during many changes that have taken place in the composition of the ocean, while the spongilla, living constantly in the same unaltered medium, has retained its primitive simplicity, it is highly probable that the vast abyss, in which the spongilla originated and left its progeny, was fresh, and has gradually

become saline, by the materials brought to it by rivers, like the salt lakes of Persia and Siberia.[60]

The assumption of an evolutionary relationship between freshwater and marine sponges here is quite explicit. It is also quite clear from this paper that Grant saw evolutionary change as being driven by the organisms' conditions of life, such as the chemical composition of seawater. Species that live in environments not subject to change can therefore be expected to remain unchanged indefinitely, as had the species of freshwater sponge. Grant then went on to say why this freshwater sponge should be considered a primitive form, based on the siliceous nature of its skeleton. He noted that 'its aptness for secreting silica, and the abundance of that earth in its skeleton, show the period of its creation to have been nearly synchronous with that of the siliceous or primitive rocks'.[61] The implication here is that these primitive creatures first came into being in an ocean rich in silica, which was in the process of precipitating out to form the crystalline primitive rocks. The silicate rocks were therefore the first to precipitate out as the ocean cooled and its chemical composition changed over time. This account accords perfectly with the Wernerian theory as taught by Jameson to his students. Grant's clear espousal of this model is strong evidence that his transformist views were integrated with a fundamentally Wernerian model of earth history.

Later the same year Grant repeated his views on the evolution of sponges in a paper on the structure of siliceous sponges published in the first number of the *Edinburgh New Philosophical Journal*. Here Grant suggested a family tree of sponges based on the form of the spines, or spicula, which make up the skeletons of many species. He traced the development of the spicula from the simple forms found in freshwater sponges through three stages of increasing complexity, first to forms where 'the unnecessary and probably hurtful embedded point has been removed', and finally to the most complex jointed spiculum.[62] Grant related these changes directly to function, as he considered that the more advanced forms were better suited for defending the sponge against predators, as 'at the time of its formation, animalicules of larger magnitude swarmed in the heated ocean.'[63] Here, Grant also made it clear that he believed that the oceans of earlier epochs had been not only different in chemical composition, but also hotter than at present. The earth had consequently experienced progressive cooling during its geological history. We therefore have, according to Grant, at least three factors in the history of the earth that could drive the evolution of life: first, there was the gradual decrease in the temperature of the earth; second, there was a change in the chemical composition of the ocean from a primordial solution highly charged with silicates and other minerals to the much more dilute seawater of the modern epoch; and third, the gradual increase in the salinity of seawater as salts were leached from continental rocks and washed into the oceans. Grant does not explicitly mention the fall in sea levels that formed a central element of Werner's theory, although the changes in the chemical composition of the oceans and their connection with the formation of the different types of rock deposited during different epochs are entirely consonant with the Wernerian model.

We have seen above that Ami Boué wrote an article on the progressive nature of the history of life, and its relationship with the Wernerian model of the history of the earth, that appeared in the *Edinburgh New Philosophical Journal* in 1826. While the picture of the history of life presented in this article is open to an evolutionary interpretation, it stops short of making any explicitly transformist claims. However, as Goulven Laurent has demonstrated, there is significant later evidence from other sources that this significant Wernerian geologist and former student of Jameson was indeed also a transformist and an admirer of the theories of Lamarck and Geoffroy.[64] His credentials as a transformist are left in little doubt by a 'résumé of the progress of geological sciences during the year 1833' that he wrote for the *Bulletin de la Société Géologique de France*. In this work he stated:

> The naturalist who restricts the circle of his ideas to the short duration of his life will necessarily be directed to the ancient idea of the species as a being *sui generis* formed once for all time, which must perpetuate itself as such, at least as long as the present laws of nature remain in effect. The authority of scholastic writings and the most ancient legislators also corroborate this opinion, engraved in the memory from the most tender infancy. On the other hand, in examining the whole scale of creations, living as well as fossil, in ignoring individual instances in order to see the whole, set in motion by a subtle material that is disseminated everywhere, one easily arrives with the Lamarcks, the Geoffroys, and other great naturalists, at an entirely different conclusion.[65]

Finally, as an illustration of just how prevalent the connection between the Wernerian model of progressive, directional change in the physical conditions on the surface of the earth and evolutionary speculation was in Edinburgh in the early nineteenth century, we turn to the writings of an avowed enemy of transformism, the extra-mural anatomy lecturer John Fletcher. In his *Rudiments of Anatomy* (1835), which was based on the lectures he gave to Edinburgh medical students, he gave the following exposition of evolutionary theories:

It has been conjectured that, in the infancy of the organic kingdom of nature, and long after the establishment of the inorganic, none but the simplest possible tribes of plants and animals, as the fungi and polypi, existed. Many of these, it is supposed, continued to be propagated by either simple division or germs; while some of them on the contrary, under the influence of different external circumstances connected with the changes to which the globe itself was gradually subjected, underwent, on the progressive supply with their aliment of fresh materials, in the forms of monades or organic molecules – the supposed nature of which will be elsewhere explained – a greater development than was consistent with the retention of their original characters, and hence resulted perhaps some higher tribes of acotyledonous plants, and among animals, the mollusca and articulata.[66]

While Fletcher went on to roundly repudiate such evolutionary fantasies, this is a model of evolution to which Grant or Jameson would surely have been happy to subscribe. We know that Fletcher was indeed acquainted with Grant and cited him regularly in his work. It is therefore no surprise that when Fletcher presented his readers with an outline of evolutionary theory in his textbook, it was to the Wernerian model, as developed and promoted by Edinburgh's Wernerian geologists and their like-minded colleagues across Europe, that he turned.

## WERNER, LAMARCK AND GEOFFROY IN EDINBURGH

The discussion in this chapter of the Wernerian influence on the development of evolutionary theories in the Edinburgh of the early nineteenth century raises a number of important questions regarding their relationship with the evolutionary theories derived from French sources that will be discussed in more detail in Chapter 5. What do the sources examined above tell us about the relationship between the theories of Lamarck, Geoffroy and Wernerian geology in Edinburgh at this time? To what extent were the theories of Edinburgh's evolutionary thinkers derived from Lamarck and Geoffroy and in what ways did they differ from them? And to what extent did they allow Wernerian geological doctrines to shape the models of evolution that were adopted? As James Secord, the historian who first properly drew attention to this important group of thinkers, has dubbed Jameson, Grant and their associates the 'Edinburgh Lamarckians', it is to Lamarck's influence I will turn first.

In examining the influence of Lamarck's theories, the anonymous article 'Observations' published in the *Edinburgh New Philosophical Journal* in 1826 is a good place to start. This paper mentions Lamarck by name and praises his theories at some length. However, the mechanism of evolution that is proposed in this article is radically different from the one suggested by Lamarck. This is because the author of the paper proposed a relationship between the history of the earth and the history of life that has no equivalent in Lamarck's writings. It is assumed throughout the paper that the conditions of existence on the earth have changed in a progressive, directional way. This is quite counter to the opinions of Lamarck, who was himself very much a uniformitarian in regard to the history of the earth.[67] In a remarkable book entitled *Hydrogéologie* (1801), Lamarck had given a detailed exposition of his own views on the history of the earth. His theory has some strong parallels with Hutton's world machine, in that it suggests that the earth is constantly being reshaped through the action of two opposing forces. But while Hutton relied on gravity and heat to drive his system, Lamarck suggested rather that two opposing gravitational forces were responsible for the constant reworking of the earth's surface. The first of these is familiar from the Huttonian model; erosion through the action of water, with the help of gravity, was constantly breaking down the rock that composed the continents and carrying it down to the sea, where it was deposited. This process acting alone, would, however, tend to plane down the continents and fill in the ocean basins until ultimately they reached a state of equilibrium at the same level and the earth became a perfect sphere covered by a universal ocean. To prevent this from happening, a countervailing force was necessary. For Hutton, this was heat. Lamarck chose a different compensating force, the gravitational attraction of the moon. This force, according to Lamarck's theory, was responsible for a continuous displacement of the waters of the oceans from east to west. This led both to the constant erosion of the eastern shores of continents and the creation of new land on the western shores. (Lamarck recognised that in reality this process would be complicated by the circulation of ocean currents, leading to a more irregular distribution of land and sea than this mechanism alone would suggest.) This process would lead the continents to chase one another around the globe through geological time, as the eastern shores of the continents were eroded away and the western shores were built up by new sediment carried into the sea by erosion. As Lamarck himself put it, 'the bed of the sea, which of necessity loses from one side what it gains from the other, has already without doubt traversed once, or even many times, all the points on the surface of the globe'.[68] Lamarck implied that the continents might have circled the globe many times; a process was in effect just as lacking any 'vestige of a beginning' or 'prospect of an end' as Hutton's model. Like Hutton, Lamarck denies that any evidence remains of the primitive composition of the earth and rejects the existence of the primitive rocks of the Wernerians:

All that it is possible to say with reason on the subject of the composite materials that are observed on the surface and in the crust of the earth, is that some are more ancient than others, that is to say, they are further from their origins. But from an ancient material or one that is very distant from its origin, to a truly *primitive* material, the distance could be infinite; so that it is lacking in reason to confuse these two objects.[69]

The constant recycling of the strata of the earth would ultimately obliterate all evidence for any truly primitive strata, just as was the case for Hutton's world machine. There is no sense of directionality in this model, and so it is unsurprising that the mode of evolution that Lamarck proposed did not require any directional change in the physical conditions of the surface of the earth to drive the development of life, but could operate against the background of a dynamic equilibrium.

In Lamarck's theory of evolution, continuous change in the conditions of life therefore could not be the main driving force of evolutionary change. Instead, evolution was a necessary outcome of the organisation of the bodies of living things themselves. He believed that only the very earliest stages in evolution were driven by fluxes of subtle fluids from their surroundings penetrating the bodies of the simplest and earliest forms of life. However, Lamarck believed that, after living things had developed beyond the most primitive monads, the main driving force behind their evolution came from fluxes of subtle fluids within their own bodies, developing and shaping their organs, and even at times bringing entirely new ones into being. For living things to increase in complexity generation by generation, they did not need continuously changing conditions. The motor of evolution was within themselves. Changes in their surroundings may certainly lead living things to modify their behaviour. This could, according to Lamarck, bring about heritable alterations in their bodies, causing them and their descendants to branch off laterally from the main line of evolutionary development. However, the gradual increase in the complexity of living things was not dependent on external factors. While possibly inspired by Lamarck, the theory elaborated on 'Observations', and indeed all the transformist texts originating in Edinburgh from the early nineteenth century that we have been looking at, are fundamentally different from his theory in this crucial regard.

The anonymous paper 'Of the changes which life has experienced on the globe' that the *Edinburgh New Philosophical Journal* published the following year also sheds much light on the relationship between the Edinburgh school of transformists and Lamarck's theories.[70] Unlike the earlier anonymous paper, it does propose a double mechanism for evolution, as does Lamarck's theory. The less import of these mechanisms is similar to Lamarck's subsidiary mode of evolution in that it was driven by unpredictable local environmental changes, which disrupted the simple pattern of development that would otherwise have prevailed. However, the main motive force from evolution came from another source, and here the article differs markedly from Lamarck, and is more in agreement with 'Observations'. There is no suggestion of a continuously acting innate tendency towards progressive change of the kind that provides the main mechanism for evolution under Lamarck's model. Instead, the main driving force determining the upward trajectory of evolution was attributed to the effect of directional change in the physical conditions, rather than an innate tendency of living things to become more perfect even in constant condistions. In this, the anonymous author seems to have substituted the Wernerian model of change in the conditions of life as the motor for evolutionary change in place of Lamarck's innate tendency of living things to become more perfect even in constant conditions.

We have the testimony of Charles Darwin as evidence of Robert Grant's admiration for Lamarck and his theories. However, when we turn to Grant's own writings, it seems that, while he may have been influenced by Lamarck, he was far from slavishly accepting his theory in its entirety. His belief that directional change in physical conditions played a role in driving the transmutation of species brought him closer both to Wernerian geology and to the evolutionary theories of Geoffroy, which also depended on continuously changing physical conditions. Grant links changes in the temperature of the earth and the chemical composition of the oceans to evolutionary change in marine invertebrates in his published work from the 1820s. Although Lamarck may have inspired some in Edinburgh to explore evolutionary explanations for the progressive history of life presented by the fossil record, the completely different mechanism for evolutionary change adopted by the Edinburgh transformists seems to have been derived principally from Wernerian geology. James Secord has remarked that: 'For the author of the "Observations", this progression of life is best explained through transmutation. Lamarck's theory is the logical consequence of Werner's.'[71] It is certainly clear that evolutionary theories such as Lamarck's were made plausible by the progressive picture of the history of life driven by directional change in the physical conditions that was offered by Wernerian geology, but only after the rejection of the principal mechanism offered by Lamarck to explain transmutation. The ultimate motor of change had to be relocated from the internal organisation of the living thing to its external conditions of life.

The driving force for the transmutation of species was not the only area in which many Edinburgh transformists seem to have disagreed with

Lamarck. We know that Lamarck believed that no species had become extinct through natural causes. This assumption was questioned by many transformists in Edinburgh. Grant, in the introductory lecture he gave at University College London on 23 October 1828, the year after he left Edinburgh, talked about 'the origin and duration of entire species, and the causes which operate towards their increase or their gradual extinction'.[72] Grant not only countenanced extinction, but also seems to have believed that catastrophes of the kind that Lamarck's great opponent, Georges Cuvier, made central to his theory of earth history were responsible for the loss of many species. In his introductory lecture he told students that: 'By thus pointing out the extensive and terrible catastrophes to which the Animal Kingdom has often been subjected, we are enabled to perceive a cause of the many apparent interruptions in the chain of existing species'.[73] The anonymous author of the 'Observations' not only suggested a causal link between evolutionary change and changing physical conditions, but also proposed a rather more Wernerian mode of extinction than the catastrophist explanation suggested by Grant. He posed the question:

> But if all living perish, may no point of duration have been fixed for the species; or do we not rather, in these signs of a former world, discover a proof that, from a change in the media in which organic creatures lived, and from powerful causes operating upon them, their power of propagation may be weakened, and at length become perfectly extinct?[74]

Jameson, the prime candidate for the author of 'Observations', certainly seems to have been convinced that some fossil species had genuinely become extinct. In a set of student's notes from his lectures in 1806, he confidently asserted that: 'Altho some naturalists will not allow it, it may be proved that several species of organic beings have become extinct'.[75] He does not seem to have changed his opinion by 1831, when he was still announcing to his students that almost all fossil animals found in strata older than the alluvial deposit were of extinct species.[76] In an undated manuscript found among Jameson's papers he speculated that the extinction of species would leave gaps to be filled by new species: 'it would seem necessary that new species should be created because other species die out – as soon as others are formed to take their place in the general oeconomy [illegible word] of nature'.[77] While Lamarck used his transformist theory as an argument against extinction, as apparently extinct species had merely transformed into new species, Jameson used the reality of extinction to argue for transformism. Without the evolution of new species, the places in the economy of nature occupied by extinct species would have to remain unfilled. This argument suggests that Jameson had a better grasp of the fossil record, the living fauna of the earth and the relationship between the two than Lamarck. It also hints at a late survival of the eighteenth-century notion of the 'chain of being' in Jameson's thinking, as he seems to imply that every place in the economy of nature must necessarily have a species to fill it.

We know that Geoffroy's rather different evolutionary theories were also known and discussed alongside those of Lamarck in early nineteenth-century Edinburgh. In the context of the present discussion of the relationship between Wernerian geology and transformism, it is worth speculating on how the prevailing Wernerian model of earth history may have influenced the reception of Geoffroy's ideas. It is evident from Geoffroy's first published evolutionary speculations in 1825 that his vision of the history of the earth was in all essentials the same as that of the Edinburgh Wernerians. He wrote: 'Physicists and geologist never at all doubt that great changes have been successively introduced in the physical conditions and the material of the globe or that these changes have greatly modified its primitive constitution.'[78] He added that '[t]o deny the influence of such circumstances on the organization [of living things] is to put oneself in the difficult position of demonstrating that such variations are impossible'.[79] Later in the 1830s Geoffroy fixed on changes in the atmosphere, and in particular on the diminution of the levels of atmospheric oxygen, as the main factor driving evolution.[80] However, in a key 1828 paper, he wrote that changes in both the temperature of the earth's surface and the constitution of the atmosphere could promote the transmutation of species.[81] The connection proposed by Geoffroy between a changing earth and a progressive history of life must have seemed obvious to those Wernerians who were sympathetic to a transformist explanation of the fossil record.

Although it is unlikely that the Edinburgh Wernerians would have accepted in full Lamarck's theory making organic evolution independent of the evolution of the earth, Geoffroy's assertion that the 'great changes that have successively been introduced in the physical and material conditions of the globe' was responsible for the great changes witnessed in the fossil record must have been extremely welcome to figures such as Jameson.[82] Natural historians writing in the 1820s often felt little obligation to acknowledge the influence of other thinkers by citing their work. Lamarck is cited only once and Geoffroy not at all in the published sources I have examined in this chapter. However, the strong agreement between the ideas of the Edinburgh transformists with those of Geoffroy, and the profound differences with those of Lamarck, suggest that they would have found Geffroy's ideas more congenial. I would contend that the influence of his key papers of 1825 on fossil crocodiles and of 1828 on the work of

François Désiré Roulin can be seen throughout the evolutionary writings of Jameson's circle in the later 1820s. The main reason for this was that Geoffroy's ideas fitted neatly into the established context of the Wernerian model of earth history, while Lamarck's did not.

We have seen how the Wernerian model of earth history, which dominated the natural history circles around Robert Jameson in the early decades of the nineteenth century, was intimately linked with a progressive history of life. The fossil record as it was known in the period provided ample evidence that species had appeared on earth in a progressive manner. This view was indeed the consensus throughout Europe, and leading English geologists such as William Buckland at Oxford and Adam Sedgwick at Cambridge would not have disagreed with it. The fundamental difference between Wernerians and catastrophists, such as Buckland and Sedgwick in England and Georges Cuvier in France, was that for Wernerians the progress of life seemed to occur by a gradual development rather than through a fresh, and more perfect, new creation after each of a series of global cataclysms. For Wernerian, new species and new higher taxa of living things came into being as soon as conditions on the earth became suitable for them. This was a consequence of the physical conditions on the surface of the earth, which had been undergoing slow, progressive changes in temperature and in the constitution of the atmosphere and oceans. The evolution of conditions on the earth was inextricably linked with the gradual appearance of new forms of life in the fossil record. This gradual, piecemeal addition of new species left the door open for transformist explanations of the patterns observed in the fossil record.

Wernerian thinkers certainly made the jump from seeing new species arising as conditions became favourable for them to accepting changes in the physical conditions as the cause of the development of new forms of life from previously existing forms. This is exactly the conclusion that several key figures in Edinburgh natural history circles came to, including Robert Jameson, Robert Grant and Ami Boué. It is possible that some of the anonymously published evolutionary articles in the *Edinburgh New Philosophical Journal* were also by otherwise unrecognised transformists, if they were not written by any of the three already mentioned. John Fleming seems to have toyed with this evolutionary explanation of the fossil record in the early 1820s, before definitively rejecting it later in the same decade. Fleming later went on to reject Wernerian geology and even the evidence for a progressive fossil record. John Fletcher, who seems to have also been well acquainted with the Wernerian-tinged model of evolution current in

Edinburgh in the 1820s also rejected it: like Fleming, probably largely on religious grounds.

The exact role played by the theories of Lamarck and Geoffroy in Edinburgh is not entirely clear. We know that eighteenth-century evolutionary theories were discussed in the lectures of John Walker, professor of natural history at Edinburgh until 1803. Walker dismissed these theories, but it is likely that the ideas stuck in the minds of some of his students, who included a young Robert Jameson. Jameson was certainly thinking about the problem of the origin of species as early as 1804, when he wrote his *System of Mineralogy*. We also know that Robert Grant was first introduced to transformism through the writings of Erasmus Darwin before he encountered the theories of Lamarck. He seems to have been aware of Darwin's transformist theories from at least his years as an undergraduate at the University of Edinburgh, as he made reference to Darwin's *Zoonomia* in an undergraduate dissertation published in 1814.[83] Much later, when he had been a professor at University College London for many years, he credited Erasmus Darwin's writings with first introducing him to the transmutation of species.[84] For Grant, at least, the theories of Erasmus Darwin proved to be the catalyst for a lifelong advocacy of the transmutation of species.

From the evidence presented in the next chapter we can say without doubt that Lamarck's theory was known by some natural historians in Scotland from at least the early 1810s. His influence seem to have grown in the early 1820s, when his *Histoire naturelle des animaux sans vertèbres* rapidly became an essential purchase for anyone with an interest in invertebrate zoology and the means to buy it. We know that Jameson owned a copy, as did the library of the Plinian Natural History Society. Almost the whole of the first volume of this work was given over to a detailed exposition of Lamarck's theory. Whether Lamarck's theories were the catalyst that led key figures in Edinburgh such as Jameson and Grant to interpret the fossil record in evolutionary terms, or whether they were well on the way to doing this already, inspired perhaps by earlier transformist thinkers such as Linnaeus and Erasmus Darwin, is a question that is impossible to answer. It is certainly true that there was a definite spike in the number of sources bearing witness to the prevalence of transformist beliefs in Edinburgh in the early and mid-1820s, immediately following the publication of the *Histoire naturelle*.

Whether or not they were inspired by Lamarck, evolutionary thinkers in Edinburgh diverged significantly from him on the question of the mechanism behind the transmutation of species. While Lamarck saw an innate tendency to progressive development as built into the organisation of living things, Edinburgh's transformists instead saw gradual, directional change in the conditions of existence on the earth as the driving force of evolution. The steady-state model of earth history proposed by Lamarck in his *Hydrogéologie*, which had much in common with that of Hutton, ruled out any general directional change in the nature of the physical conditions of the earth. For Lamarck, and for any natural historians who accepted his theory in its entirety, it was therefore strictly inadmissible as a mechanism for evolution. The Edinburgh transformists also generally differed with him on the subject of extinction, which they believed to be a real phenomenon rather than just an artefact of our imperfect knowledge, as Lamarck claimed. In consequence, they can only be described as disciples of Lamarck in the loosest sense. Their belief in the efficacy of changes in the conditions of life in bringing about evolutionary change brought the Edinburgh transformists closer to the theories of his compatriot and colleague at the Muséum d'Histoire Naturelle in Paris, Étienne Geoffroy Saint-Hilaire. Like the Wernerian geologists, Geoffroy believed in slow, progressive change in the physical conditions of the earth. In particular he emphasised changes in the atmosphere of the earth as playing the key role in bringing about evolutionary change. Geoffroy published his first openly transformist article in 1825, although it is possible that some Edinburgh figures, such as Grant, who knew Geoffroy personally, may have been aware of his views some time before then. It is entirely possible that the publication of a clutch of evolutionary articles in the *Edinburgh New Philosophical Journal* in the years 1826–9 was more than a coincidence, and that these were directly inspired by Geoffroy.

In conclusion, it would seem that Wernerian geology merged seamlessly with a transformist interpretation of the fossil record and the history of life in the minds of an important group of Edinburgh natural historians. Reading the works of Lamarck may have encouraged and inspired them, but their views on the efficacy of progressive change in the physical conditions of life in driving evolution meant that they could never accept his system in its entirety, but only in a drastically modified form. Geoffroy's model of evolution, although it was only widely known in the second half of the 1820s, provided a much better fit with the Wernerian-inflected views of Edinburgh's evolutionary thinkers. Although Geoffroy's published theories were more elaborate than anything in the sources we have from Scotland, crucially they share the same emphasis on the central importance of changes in the physical conditions of the earth. Although both Geoffroy and Lamarck without doubt had an influence on the development of evolutionary ideas in Edinburgh, the single most important factor in facilitating the acceptance and development of evolutionary interpretations of the history of life was the prevalence of Wernerian geology in natural history circles in the city and the tireless promotion of it by Robert Jameson.

## NOTES

1. Fleming, 'Natural science', in Fleming, *Inauguration of the New College of the Free Church*, p.216.
2. [Jameson], 'New publications received' (1845), p.186.
3. [Jameson], 'New publications received' (1846), p.400.
4. Brewster, Review of [Chambers'] *Vestiges*.
5. Jameson, *System of Mineralogy*, vol.1, pp.xix–xx.
6. Jameson, *Elements of Geognosy*, p.73.
7. Laudan, *From Mineralogy to Geology*, p.181.
8. Jameson, *Elements of Geognosy*, p.77.
9. Scottish Universities Commission, 'Syllabus of lectures on natural history', in Scottish Universities Commission, 'Appendix to No. XIII', *Returns, Papers, and Examinations*, p.11.
10. Jameson, Notes on natural history lectures (1830), f.5.
11. [Jameson] Note on the verso of a card bearing the inscription 'Civis Bibliotecae Academiae Edinburgenae'.
12. Laudan, *From Mineralogy to Geology*, p.87.
13. See, for example, the very full account of the neptunian theory in the set of notes taken in Jameson, Notes on natural history lectures (1835/6), vol.3.
14. Rudwick, *Bursting the Limits of Time*, p.173.
15. Hutton, 'Theory of the earth', p.304.
16. Hall, *Account of a Series of Experiments*.
17. Scottish Universities Commission, *Report Relative to the University of Edinburgh*, p.90.
18. O'Connor, *The Earth on Show*, p.47.
19. Ogilvy, 'On the Huttonian and Neptunian theories of the Earth', f.238.
20. Ibid., f.248.
21. Gilby, 'Comparative merits of the Huttonian and Wernerian theories of the Earth', f.449.
22. McDowell, 'Examination of the igneous or Huttonian theory of the earth', f.329.
23. Watson, 'Essay on petrifactions', f.225.
24. Rudwick, *Bursting the Limits of Time*, p.170.
25. Playfair, *Illustrations of the Huttonian Theory of the Earth*, p.119.
26. Rudwick, *Worlds before Adam*, p.49.
27. [Jameson], Notes on natural history lectures (watermark 1813/14), f.255.
28. Jameson, *Elements of Geognosy*, p.82. It is worth noting that Fleming quotes part of this passage in his 1850 inaugural lecture as proof of Jameson's adherence to transformism.

29. Ogilvy, 'On the Huttonian and Neptunian Theories of the Earth', f.238.
30. Fleming, *Philosophy of Zoology*, Vol.2, p.97.
31. Ibid., p.104.
32. Boué, 'Geological observation', p.90.
33. Esmark, 'Remarks tending to explain the geological history of the Earth', pp.120–1.
34. [Jameson], Notes on natural history lectures (watermark 1813/14), ff.236–8.
35. Anon, 'Remarks on the ancient flora of the Earth', p.127.
36. De Candolle, 'On the history of fossil vegetables', pp.98–9.
37. Cunningham, 'On the geology of the Lothians', p.9.
38. Fleming, *Lithology of Edinburgh*, p.15.
39. Fleming, *Philosophy of Zoology*, vol.1, p.27.
40. Ibid., p.27.
41. Corsi, 'The revolutions of evolution', p.17.
42. Boué, 'Autobiographie pour mes amis', p.ii.
43. Anon, 'Nature and importance of geology', p.299.
44. Ibid., p.298.
45. Cuvier, *Essay on the Theory of the Earth*, 5th edn, p.vi.
46. Ibid., p.431.
47. [Jameson], Untitled manuscript on the transmutation of species, ff.1–2.
48. Ibid., f.4.
49. Ibid., f.5.
50. Walker, Notes on natural history lectures (1797), f.135.
51. Desmond, *Politics of Evolution*, p.446.
52. Anon, 'Of the changes which life has experienced on the globe', p.299.
53. In an appendix to his autobiography he admitted that in 1820 he was 'extremely plutonist in comparison to my master Professor Jameson'. Boué, 'Autobiographie pour mes amis', p.ii.
54. Anon, 'Of the changes which life has experienced on the globe', p.298.
55. Ibid., pp.298–9.
56. Ibid., p.299.
57. Ibid., p.300.
58. Ibid., pp.299–300.
59. Ibid., pp.300–1.
60. Grant, 'On the structure and nature of the Spongilla friabilis'.
61. Ibid., p.284.
62. Ibid., p.350.
63. Grant, 'Observations on the structure of some silicious sponges', p.350.
64. Laurent, 'Ami Boué'.
65. Boué, *Bulletin de la Société Géologique de France: Résumé des Progrès de Sciences Géologiques pendant l'année 1833*, pp.113–14.
66. Ibid., pp.12–13.
67. See Lamarck, *Hydrogéologie* and Burkhardt, *Spirit of System*, p.111.
68. Lamarck, *Hydrogéologie*, p.68.

69. Ibid., p.139.
70. Anon, 'Of the changes which life has experienced on the globe'.
71. Secord, 'Edinburgh Lamarckians: Robert Jameson and Robert E. Grant', p.9.
72. Grant, *Essay on the Study of the Animal Kingdom*, p.6.
73. Ibid.
74. Anon, 'Nature and importance of geology', p.298.
75. Jameson, Notes on natural history lectures (1806), f.105.
76. Jameson, Notes on natural history lectures (1830/1), ff.134–5.
77. [Jameson], Untitled manuscript on the transmutation of species, f.6
78. Geoffroy, 'Recherches sur l'organisation des gavials', p.150.
79. Ibid., p.150.
80. Geoffroy, *Études progressives d'un naturaliste pendant les années 1834 et 1835*, p.119.
81. Geoffroy and Serres, 'Rapport fait à l'Académie royale des Sciences sur une mémoire de M. Roulin', p.207.
82. Geoffroy, 'Recherches sur l'organisation des gavials', p.151.
83. Grant, *Dissertatio Physiologica Inauguralis*, p.8.
84. Grant, *Tabular View*, p.v.

# 5

## Edinburgh and Paris

Into a climate favourable to transformist speculations created by Robert Jameson and his Wernerian geological doctrines came a flood of new theories from continental sources on the nature and origin of species from the early 1820s onwards. This was hardly a surprising development, as Edinburgh had a long tradition of openness to continental ideas and maintained strong links with scholars across Europe. Jameson himself had studied with Werner in Freiberg and both Knox and Grant attended Geoffroy's lectures at the Muséum d'Histoire Naturelle in Paris, where they absorbed the latest theories in that great centre of innovation in the natural sciences. Although there is some evidence that the theories of German thinkers such as Johann Friedrich Meckel (1781–1833), Lorenz Oken (1779–1851) and Friedrich Tiedemann (1781–1861) were known in Edinburgh in the 1820s, references to them in the written sources are comparatively rare. For the most part the transformist ideas discussed in Edinburgh seem to have originated in the Jardin des Plantes in Paris, where the Muséum d'Histoire Naturelle was the home of two of the most important transformist thinkers in early nineteenth-century Europe, Jean-Baptiste Lamarck and Étienne Geoffroy Saint-Hilaire.

In this chapter I will be exploring the impact in Scotland of Lamarck and Geoffroy. First I will look in more detail at the theories of these two important transformist thinkers. Then I will examine the reception of their ideas among advocates of transformist theories, among those who rejected them, and also among the significant group of commentators who reserved judgement. I will start with the reception of the works of Lamarck, who was by a considerable margin the most cited transformist thinker in Edinburgh in the 1820s and 1830s. I will then turn to Geoffroy, whose transcendental anatomy had an enormous influence on early nineteenth-century natural history, both through his published work and his role as the teacher of some of the key figures in Edinburgh natural history. In addition to assessing the reception of Lamarck and Geoffroy, I will go on to explore how their ideas were adapted and modified by Scottish transformist thinkers when developing their own theories.

## CONTEMPORARY TRANSFORMISM IN FRANCE: JEAN-BAPTISTE LAMARCK AND ÉTIENNE GEOFFROY SAINT-HILAIRE

Before examining Lamarck's mature theories of the transmutation of species, it is important to take a look at how his ideas were underpinned by his understanding of the meanings of 'nature' and 'life'. In *Histoire naturelle*, Lamarck defines nature as 'an order of things, alien to matter, determinable through the observation of bodies, and of which the whole constitutes in its essence an inalterable power, constrained in all its actions, and constantly acting in all parts of the universe'.[1] Based on this description, it seems clear that the Newtonian laws of gravitation would have provided an ideal illustration of Lamarck's nature in action, and this may well have been the model he had in mind. Earlier in the same work he discussed the operation of nature on living things, describing it as 'the power, in some manner mechanical, which has given existence to the diversity of animals, and of necessity made them what they are'.[2] Living things were therefore subject to unchanging natural laws in the same way that the planets were constrained in their orbits by the laws of gravitation. Lamarck made it clear that nature was not an intelligence, but merely applied mechanically the natural laws established by God: 'an intermediary between GOD and the physical parts of the universe, for the execution of the divine will'.[3]

We turn now to Lamarck's concept of organic life. As he stated in his *Histoire naturelle*, for Lamarck, 'every action or phenomenon observed in a living body is at the same time a physical action or phenomenon, and the product of organisation'.[4] This conception of the nature of life is fundamentally opposed to vitalism; life is not the manifestation of an immaterial essence, but rather the product of the organisation of matter acting in accordance with essentially mechanical laws of nature. It is therefore a radically materialist theory of life. Lamarck goes on to explain how nature goes about organising non-living matter to create life. This process is directly analogous to the normal process of fertilisation that takes place during sexual reproduction. Lamarck believed that a 'subtle vapour' emanating from the seminal fluid acted as an organising principle that transformed the ovum into a living being. Supplied with the right materials by nature, Lamarck believed that 'subtle fluids', principally heat or electricity, present in the environment, could generate life from non-living matter. As he himself put it:

Why may not heat and electricity which, in certain parts of the world and during certain seasons are so abundant in nature, above all at the surface of the earth, act on certain substances in a favourable state and circumstances which it finds there, and perform that which the subtle fertilising vapour performs on the embryos which it organises and prepares to enjoy life?[5]

In his *Organisation des corps vivants* the subtle fluids were given as heat, which he usually referred to in the guise of 'caloric', a fluid believed to embody heat, and electricity. In his *Histoire naturelle* twelve years later he included the magnetic fluid and possibly light to the list, but the basic theory remained the same.[6] Indeed Lamarck considered that all of these fluids might simply be different states of the same substance.[7] These fluids were 'subtle' because they were *incontenable*, in other words they could not be contained by surrounding matter, but could flow through it. They were considered to be so tenuous that they could not be perceived directly by the senses, but only known from their effects.[8] One of these effects was manifested in their ability to act on *contenable* fluids, such as water or the bodily fluids of living things. The rotation of the earth on its axis and its movement around the sun caused continual fluxes of subtle fluids on the surface of the planet, the action of which on suitable matter could bring life into being.[9] It was the action of these subtle fluids on tiny, spontaneously generated gelatinous bodies that transformed them into the simplest living beings. In the *Organisation des corps vivants* Lamarck explained in more detail how the organisation of living beings comes about: 'In this way the uncontainable fluids trace the first traits of the simplest organisation, and then the containable fluids develop it by their movements and other influences'.[10] In this way the simplest forms of life were, and continued to be, produced from inanimate matter through the operation of natural laws. These first 'sketches of life', as Lamarck calls them, were therefore at the same time both the oldest and the youngest of living things. Lamarck believed that, as the most simple organisms found in nature lived in water, 'it can be regarded as a fact that it is uniquely in water that the animal kingdom had its origin'.[11]

These 'living points' had no organs, could not move independently, and relied for nutrition on the deposition of materials brought into their bodies by the fluxes of *contenable* fluids, such as water. These fluxes were driven ultimately by the movements of the subtle fluids which surrounded them in their environment. However, the process of development did not stop at this point, but continued to complicate the structures of these simplest of living things. The earliest organ to appear was the digestive tract, at first with only one opening, as in polyps. As living things became more complex, the fluxes of subtle fluids became internalised, so that the further development of the organism was less dependent on external influences, and living things were less at the mercy of the environment for their existence. Further development then brought other organs into being in a step-wise manner. With these new organs came new faculties, which could not exist without the organs appropriate to them. Although Lamarck never worked out a mechanism for heredity, he believed that changes that occurred during the lifetime of an organism were transmitted to its offspring, and so the upward trend of development continued generation after generation. All things being equal, organisms would tend to ascend through a fixed series of increasingly complex forms over time. However, according to Lamarck, all things were not equal, as the environment had a strong effect on the development of the bodies of animals. Changes induced by their conditions of life would tend to push animals off the fixed developmental track they would otherwise be constrained to follow. As Lamarck puts it in his *Philosophie zoologique*:

> It will, in effect, be evident that the state in which we see all animals is, on one hand, the product of the increasing *composition* of their organisation, which tends to form a regular gradation, and, on the other, the product of a multitude of very different circumstances, which tend continually to destroy the regularity of the gradations of their increasing composition of organisation.[12]

Lamarck's theory turns the argument for design on its head, since animals are not designed to suit their environments, but rather their environments, and the needs that they find themselves subject to, have moulded them into what they were. As Lamarck puts it in the *Organisation des corps vivants*:

> It is not organs, that is to say the nature and forms of the parts of the animal body, which have given rise to its habits and faculties; but, on the contrary, the habits, way of life and circumstances which its ancestors have encountered which have over time determined the form of its body, the number and state of its organs, and finally the faculties of which it enjoys the use.[13]

Thus new needs, brought about either by the increasing level of organisation of the organism or by environmental change, produce new habits, which in turn lead to the development of existing organs or the production of new ones to meet that need. Conversely, organs that are no longer required and consequently no longer used will diminish and ultimately disappear. This all takes place in an entirely mechanical fashion through the flow of fluids to the appropriate organ, stimulating its growth and development. Dismissing the term 'inheritance of acquired characteristics' often used by later commentators, but never by Lamarck himself, Corsi points out that what are inherited are not physical characteristics but biological processes; 'it was not "characters" that were acquired during the lifetime of the organism, but only a higher or lower degree of organic fluid flow or, in general, a small difference in fluid distribution patterns'.[14]

Lamarck's belief in the mutability of species led him to propose that they did not exist in nature outside the context of a particular moment in geological time; they 'only have a relative permanence, and are only temporarily invariable'.[15] Species only seemed stable over the relatively short period of human history; in the much longer time frame of geological time, this apparent stability would disappear. The classification of living things into species by naturalists was therefore entirely artificial. Nonetheless, the species concept was useful in the study of living things and Lamarck did not propose abandoning it altogether. He suggested that 'it was useful to give the name "species" to all collections of similar individuals, which reproduction has perpetuated in the same state as long as their circumstances do not change enough to vary their habits, their character and their form'.[16] Classification therefore became simply a tool for the naturalist, but did not reveal the true order of nature, which could only be understood as a process unfolding over vast spans of time. In a memorable passage from a memoir published in the annals of the Muséum for 1802, Lamarck compared natural historians who did not take account of the vastness of the history of the earth to a group of insects living in a building:

> I seem to hear some of those little insects who live for only a year, living in some corner of a building, and who one can imagine busy among themselves consulting their traditions to pronounce on the age of the building in which they find themselves; going back through their paltry history as far as the twenty-fifth generation, they unanimously decide that the building that shelters them is eternal and that it has always existed, as it has always appeared the same to them, and they have never heard it said that it had any beginning.[17]

Lamarck's theory also has important implications for the overall shape of the history of life on earth. First, he considered that animals and plants had entirely separate origins; both owed their origins to fluxes of subtle fluids acting on matter, but differed in the nature of the matter in the two cases. If the tiny mass of matter into which organisation was to be introduced was gelatinous, it would become an animal, if it was mucilaginous, it would become a plant.[18] For this reason plants lacked the fundamental property of irritability possessed by all forms of animal life.[19] In consequence Lamarck denied absolutely the existence of intermediate plant-animals, or zoophytes, which many natural historians viewed as a link between the animal and vegetable kingdoms.

Not only did Lamarck consider that plants and animals had separate origins, but he also envisaged that the animal kingdom could be classified into two main branches with separate origins. One, consisting of the infusoria, polyps and radiaria (and in later versions also the acephala and the mollusca) were ultimately descended from the first 'living points', which had come into being through spontaneous generation from non-living matter. The second branch, consisting of all the remaining invertebrates, including insects, crustaceans and annelids, owed their origin to the spontaneous generation of parasitic worms within the bodies of the animals belonging to the other, older branch.[20] This shown clearly in the diagram from Figure 5.1, from *Histoire naturelle des animaux sans vertèbres*, which illustrates the two main branches of the invertebrates. The relationship of the vertebrates is left somewhat ambiguous in this diagram. In the earlier diagram from the *Philosophie zoologique*, shown in Figure 5.2, Lamarck had made the position of the vertebrates clearer. This reflects changes in Lamarck's thinking

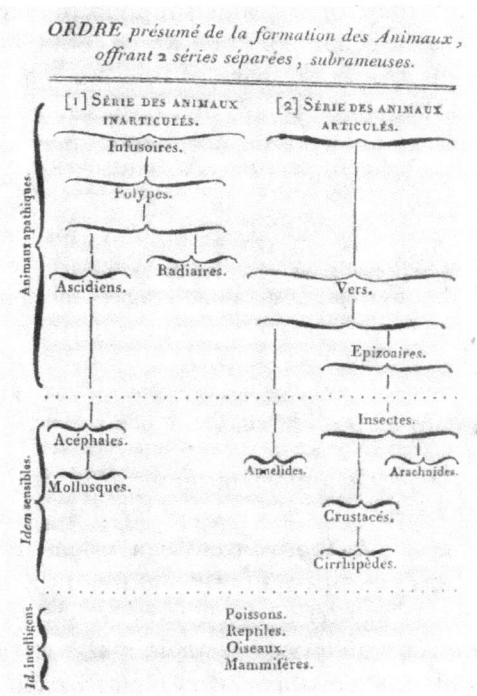

**Figure 5.1** 'Presumed order of the formation of animals in two separate series.' From Lamarck's *Histoire naturelles des animaux sans vertèbres*, vol.1, p.457. Image from the Biodiversity Heritage Library. Digitised by Smithsonian Libraries.www.biodiversitylibrary.org.

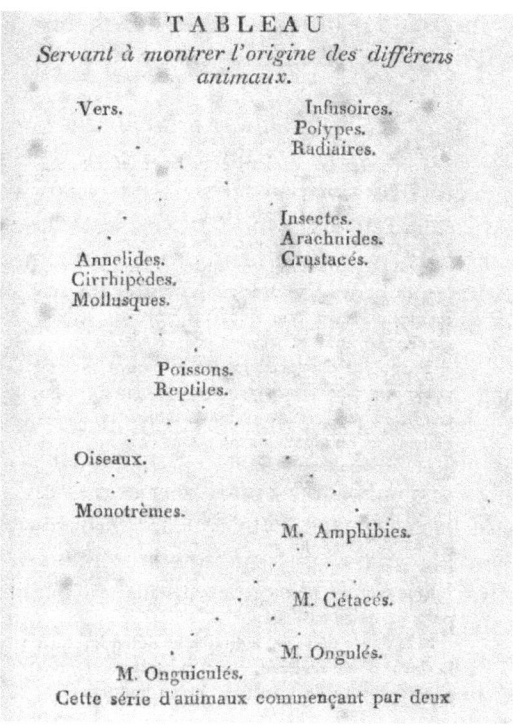

**Figure 5.2** 'The origins of the main subdivisions of the animal kingdom.' From Lamarck's *Philosophie zoologique*, vol.2, p.463. Credit: Bibliothèques de l'Université de Strasbourg (document BNU en dépôt).

between 1809 and 1815. Figure 5.2 embodies Lamarck's ideas as they stood in 1809, when he published his *Philosophie zoologique*. Here the vertebrates are clearly shown as part of the series descending from the parasitic worms. However, by 1815 he had become uncertain of where the connection lay between invertebrates and vertebrates, and could only conclude that 'this transition is still unknown'.[21] Among other less significant changes, Lamarck also changed his mind about the place of the molluscs between 1809 and 1815, moving them from the second to the first series. However, despite these changes, the overall pattern remained largely unchanged.

It is tempting to view the diagrams in Figures 5.1 and 5.2 as genealogical trees. However, as Lamarck believed that new animals were continuously coming into being through spontaneous generation even at the present time, which then proceeded through the series of stages of Lamarck's system, it would be more accurate to think of them as developmental pathways that have been followed, and continue to be followed, by all animal forms as they pass through the different levels of organisation. The branching pattern resulted from the influence of environmental factors modifying the series.

It is noteworthy that there is an almost total absence of evidence drawn from the fossil record in Lamarck's major theoretical works. This is all the more surprising given the care he had taken in his earlier work in conchology to identify fossil 'analogues' for modern species.[22] In a series of papers in the *Annales* of the Muséum in the first decade of the nineteenth century, as well as in his later taxonomic works on invertebrates, he attempted to identify the living 'analogues' of fossil species. As early as 1799 Lamarck had stated his belief that 'it is absolutely essential to research and determine the living or marine analogues of the great number of fossil shellfish which are found buried in the middle of our vast continents'.[23] The existence of such analogies was, for Lamarck, strong evidence against the reality of the periodic catastrophes and subsequent mass extinctions envisaged by Cuvier. That some of these living species, while clearly still showing an identity with their fossil equivalents, had undergone minor changes in form was the only explanation for Lamarck; in a paper published in the *Annales* of the Muséum in 1802, he noted that he suspected that the fossil mollusc species *Voluta musicalis* 'was the analogue of the *volute musica* of Linnaeus, a little changed in the course of time'.[24] Jordanova has commented that fossils 'were certainly an important element in the *genesis* of his historical approach to nature, but a detailed examination of the fossil record had no place in his arguments for transformism'.[25] Corsi has explained this as a result of Lamarck's rejection of extinction, suggesting that:

> Lamarck never relied on paleontological data, convinced as he was that all fossils, with a few exceptions (essentially the remains of animals destroyed by man), were still alive somewhere on Earth or at the bottom of the seas. For Lamarck, as for a number of his followers, the beautiful fossil ammonites found embedded in rocks were still thriving in the oceans.[26]

Although there may be some merit in this argument, it is also true that Lamarck argued that, although species did not become extinct, they might be drastically transformed over time; although the direct ancestors of the ammonites might still have been with us, they might not be readily recognisable as such after such a great span of time. I would argue rather that Lamarck's failure to use evidence from the fossil record is entirely in keeping with his style of working, which owes much to the eighteenth-century tradition of Buffon, Lamarck's patron in the early part of his career, and to whom he owed his position at the Muséum d'Histoire Naturelle. This tradition, far removed from the inductive methodology that was coming to dominate science in the early decades of the nineteenth century, exhibited a profound faith in the power of reason alone, working from solid premises, to resolve the mysteries of the natural world.

While Lamarck was promoting and elaborating this theory, one of his younger colleagues at the Muséum, Étienne Geoffroy Saint-Hilaire, was developing his own, rather different, transformist theory. It is to him that we turn next. Geoffroy published the first volume of what was to be his most influential work, his *Anatomie Philosophique*, in 1818. In this work he stated that 'it is fairly easy to reduce to the unity of composition the diverse forms of organization of the vertebrates'.[27] His belief that the anatomies of all vertebrates conformed to a single ideal plan led him to reject the idea of a single scale of being. While comparative anatomy had traditionally made the human body its point of reference for the study of the 'lower' animals, which could be arranged in a series of decreasing perfection from the human form, Geoffroy's theories led him 'not to give preference to any anatomy in particular, but to consider the organs first where they are at the *maximum* of their development, in order to follow them step by step to the zero of their existence'.[28] This was a radical departure from conventional notions of the relationships between different animal groups, including those of his colleague at the Muséum, Lamarck.

Between the publication dates of the two volumes of the *Anatomie Philosophique*, Geoffroy announced a radical extension of his theory of unity of plan to insects in his lecture course at the faculty of Sciences in 1820. For Geoffroy, the exoskeleton of the insect was homologous with the spinal column of vertebrates. The internal organs of insects were therefore enclosed within their spines. As Geoffroy put it: 'In the last analysis, we arrive at this result: every animal lives inside or outside its vertebral column.'[29] Geoffroy had already embraced the theory of Étienne Serres (1786–1868) that during embryonic development higher vertebrates passed consecutively through forms that represented the adult forms of each of the main divisions of the lower vertebrates during foetal development. He now extended this theory to include insects: 'insects occupy a place in the series of the ages and of the developments of the higher vertebrates, that is to say, that they actualise one of the conditions of their embryo, as the fishes do for one of those of their fetal stages.'[30] In 1830, a paper was submitted to the Academy of Sciences by Pierre Stanislas Meyranx and M. Laurencet, the latter being a man so obscure that his first name is now unknown. This paper suggested that the bodies of molluscs, specifically cephalopods, also partook of the same unity of plan as vertebrates.[31] Geoffroy was delighted with this further extension of the principle of unity of plan, although it proved to be highly controversial.

There is no evidence to suggest that Geoffroy was a transformist prior to 1825, when he published his first explicitly transformist paper, although this does not prove that he did not toy with the idea earlier.[32] It is not too difficult to see how a belief in a common body plan for all animals could lead to the idea of common descent. Comparative anatomy, was, however, only one of the four main pillars of Geoffroy's transformism. The others were Serres' embryology, the evidence of vertebrate palaeontology and Geoffroy's own experiments in teratology, the study of congenital abnormalities, or 'monstrosities' as they were known in the early nineteenth century.

Right from his first paper on the subject in 1825, 'On the question of whether the gavials, today spread throughout the eastern parts of Asia, descend, by way of uninterrupted generation, from the antediluvian gavials',[33] Geoffroy's theories were grounded in palaeontological discoveries. The principal fossil evidence used by Geoffroy in support of his theories was the remains of fossil reptiles found at Caen, Le Havre and Honfleur. In this key paper, Geoffroy suggested that these fossil reptiles were the ancestors of modern crocodiles. Cuvier had used the apparent identity of the Ancient Egyptian mummified crocodiles brought back from Napoleon's expedition to Egypt in 1798–1801 with modern species as evidence against transformism. However, like Lamarck, Geoffroy considered that the 'several thousands of years that had flowed by since the earth took its present form do not constitute a sufficiently long lapse of time to have brought about significant and permanent variations in the organisation of living beings'.[34] Nonetheless, he went on to claim that, unlike Cuvier, he had been able to detect differences between the modern and the mummified crocodiles that, although slight, provided evidence of an evolutionary relationship between the fossil reptile Teleosaurus and modern crocodiles, since 'the points of variation which I believe I have found there relate to the organic system in which also resides the differences between Teleosaurus and the crocodiles'.[35] In other words, the mummified crocodiles represented an intermediate form between modern crocodiles and the fossil Teleosaurus, while being very much closer to the former, due to the relatively short time that had elapsed since the mummified crocodiles had been alive.

In an 1828 report to the Royal Academy of Sciences co-authored with Serres on a paper on modifications observed among domestic animals transported to different environments by François Désiré Roulin (1796–1874), Geoffroy developed his ideas considerably beyond the relationship between Teleosaurus and modern crocodiles. He also left Roulin's own relatively modest claims far behind, asserting that Roulin's research 'leads to an understanding of the way in which extinct animals are, through uninterrupted generations and successive modifications, the ancestors of the animals of the present world'.[36] He also went so far as to propose the following 'progressive series': 'Icthyosaurus, Plesiosaurus, Pterodactylus, Mososaurus, Teleosaurus, Megalonix, Megatherium, Anoplotherium, Paleotherium, etc'.[37] Teleosaurus, meaning 'completed lizard', was so named by Geoffroy because he considered it a transitional form, showing mammalian characteristics; in the series above it appropriately forms the link between reptiles and mammals.

Fossil bones provided evidence for the transmutation of species, but Geoffroy also developed a detailed theory to explain the mechanism that brought it about. Teratology played an important role in this theory. Geoffroy believed that the same forces that led to the production of monstrosities were those that drove the transmutation of species. As he put it in his 1825 paper on fossil reptiles: 'That which, in the great operations of nature, demands a considerable span of time, is nevertheless accessible to our senses, and can be found reproduced in miniature and under our eyes in the spectacle of monstrosities, whether produced accidentally or deliberately'.[38] According to Geoffroy, these changes were brought about by modifications in the environmental conditions experienced by animals in the course of their foetal development. In order to test these ideas, he resorted to what one historian has aptly described as 'experimental transformism'.[39] These experiments were conducted on chicken eggs in a hatchery in the village of Auteuil. Believing that the constitution of the atmosphere had a profound effect on development, Geoffroy attempted to provoke monstrosities by varying the exposure of the foetus to atmospheric gases; for example he would file or prick the shell to facilitate the entry of gases, or coat parts of the shell in wax to reduce it. Although the results were, unsurprisingly, inconclusive, Geoffroy claimed to have had some success in producing monstrosities, reporting that: 'I made monsters at will, and even better, it is clear that by varying my procedure, and through the success of various attempts and trial and error, I was able to produce them with one quality or another'.[40]

Geoffroy believed that the constitution of the atmosphere had changed gradually over geological time. This change played a double role in the history of life; on the one hand it provoked the production of mutations, on the other it gradually made the earth uninhabitable for older forms, which were replaced by new ones arising from the mutations that these changes themselves generated. Of course, not all mutations were favourable, and many of the 'monsters' produced would not survive. The changing environment therefore imposed a selective pressure, which determined the forms that would survive and flourish:

The imperceptible changes from one century to the next end up accumulating and reaching a certain point at which respiration becomes difficult and finally impossible for certain organ systems: it therefore requires and creates for itself another arrangement, perfecting or altering the pulmonary cells in which it operates; modifications which may be beneficial or harmful. These then propagate and spread through the rest of the animal economy. If those modifications lead to harmful effects, the animals which undergo them will cease to exist, to be replaced by others, with forms modified to suit the new circumstances.[41]

In an article in the *Revue Encyclopédique* of 1833, Geoffroy made clear that 'the decreasing quantity of oxygen relative to the other components of the atmosphere' is the most important factor.[42] He explained why the oxygen content of the air had been decreasing over geological time through recourse to a geo-chemical explanation:

the imagination cannot fail to be frightened by the prodigious volume of shells produced since the first appearance of molluscs and by the thickness that their remains has added to the crust of the globe. Yet, in the final analysis, all the shells are reduced to insignificance by the calcium saturated in the principle of combustion, that is to say chalk, earth, in great part formed of fixed oxygen.[43]

The oxygen lost to the atmosphere was therefore locked up over geological time in rocks in the form of calcium carbonate (rather picturesquely designated as 'calcium saturé du principe comburant' by Geoffroy), the production of which in the oceans of the world was a gradual but continuous process.

Opposing the causes of variation in animals was a power called the *nisus formativus*, a concept, at least as far as its role in reproduction is concerned, that played a similar role to Buffon's internal mould. Geoffroy defined this as 'the principle which presides over the successive order of generations, which compels the return to the same forms, and in consequence the reappearance of the same species, that is to say the tendency to regular development'.[44] So the history of life on earth could be explained by the struggle between these two competing forces; the *nisus formativus* promoting stability and the physical conditions compelling change. Geoffroy's transformism therefore shared with Lamarck the idea of two competing principles. The major difference was that in Lamarck's thought both principles promoted change, although in different directions. For Geoffroy the two principles were antagonistic to one another: one promoting change, the other stability. Lamarck's transformism was driven by an innate tendency towards perfection that was built into the living organism; for Geoffroy, progressive change was entirely the result of external factors acting on living things.

Although they were both transformists, Geoffroy's theory differs in other significant ways from Lamarck's, both in scope and in its details. Unlike Lamarck, Geoffroy does not concern himself with the origins of life. He shows no apparent interest in the question of spontaneous generation, and addresses only how already existing organisms are modified by their changing environment. This has led Goulven Laurent to suggest that for Geoffroy, 'nature does not start, as for Lamarck, by creating very simple animals, in order to then progressively complicate them, but she commences with a "plan of construction", with an "ideal being"'.[45] Unlike Lamarck, Geoffroy's model proposed a direct role for environmental influences acting on animals in the course of foetal development, while Lamarck saw them as acting indirectly through the development of new habits after birth.[46] Their very different approaches to transformism may be in part be explained by the significant age difference between the two men. Lamarck had been trained in the philosophical, system-building tradition of Buffon, who had been his patron at the Museum of Natural History in his early days there, while Geoffroy was of the same generation as Cuvier, and therefore imbued with a more inductive, experimental approach.

Appel has rightly remarked that Geoffroy's theories were not as comprehensive as Lamarck's.[47] The more limited scope of Geoffroy's transformism may have made it more acceptable to some contemporaries than Lamarck's all-encompassing system building. Lamarck's theory-heavy but evidence-poor system looked increasingly like a throwback to eighteenth-century models of scientific practice. The new inductive model was intensively promoted in France by the immensely influential Cuvier, who deeply distrusted theory in general and was violently hostile to transformism in particular. Unlike Lamarck's theories, Geoffroy's were based on detailed anatomical research and experimentation, which made his theories harder to dismiss as armchair speculation. Despite this, Geoffroy was determined to defend his philosophical approach to scientific practice against what he saw as Cuvier's advocacy of sterile fact collection. In two of his published works he used the following parable to illustrate the importance of theory over dry fact gathering; once in a memoir read to the Academy of Sciences in 1831, and again in *Études Progressives d'un Naturaliste pendant les Années 1834 et 1835* in 1835.

> Paul has the desire and the means to procure all the pleasures of life: he is intelligent, inventive, and he applies himself to find and gather together that which he supposes ought to be necessary for him. He stocks his cellar with the best wines, he fills his woodshed with all the wood necessary to keep him warm: he acts with the same discernment in regard to all the other probable items for his consumption. They are chosen for their good quality and conveniently arranged

and a wise order reigns everywhere. But when he has achieved this, Paul stops. He will not drink this wine, he will not warm himself with this wood, he won't use any of the items he has assembled. – *But*, you will say to me, *your Paul is a madman.* – I agree.[48]

## LAMARCK IN SCOTLAND

While we know that seventeen Scottish students took Lamarck's course in invertebrate zoology between 1813 and 1823, neither Knox nor Grant was among them.[49] As far as we know, none of Lamarck's Scottish students went on to embrace transformist views on their return. Those who accepted at least some elements of his transformist theories will therefore have learned about them either from his published work or at second hand. Lamarck's first major transformist work, his *Philosophie zoologique*, received rather little attention in Scotland on its first publication, although it did elicit one important review from a Scottish natural historian.[50] It was only after the publication of his *Histoire naturelle des animaux sans vertèbres* in seven volumes between 1815 and 1822 that references to Lamarck's theories dramatically increased. Almost the entire first volume of this work was taken up by a detailed exposition of Lamarck's transformist speculations.

The first known published reference to Lamarck's transformist theories by a Scottish natural historian appeared in a review of Lamarck's *Philosophie zoologique* in the *Monthly Review* of August 1811. This was written by Lockhart Muirhead (1765–1829), the University of Glasgow's regius professor of natural history. It was also the earliest known review of Lamarck's magnum opus in any English-language periodical. Muirhead was happy to accept that species could be modified by the effects of physical conditions of life or through domestication. He noted:

> A change of climate, for example, or domestication, or intercourse between animals nearly allied in physical constitution, are observed to alter the forms and habits of the progeny; and to induce and multiply such changes on the original diversities, that we feel ourselves lost among breeds, species, and varieties.[51]

Given that Lamarck had made a convincing case that such transmutations were possible, Muirhead found nothing fundamentally implausible about Lamarck's theories. Nor did he find them problematical on religious grounds. Whether God had chosen to create living things through a single, supernatural act of creation or relied on secondary causes to bring species into existence through a gradual development did not seem to him a question that should throw conventional religion into doubt. For him:

The answer is obvious, that the supreme Creator may educe his works in the way and manner which he deems most fit; that he may bid all tribes of beings spring into existence at once; or may so constitute and indue the most rude and simple forms, that they shall prove the proximate sources of gradually unfolding varieties[.][52]

For Muirhead, the problem was lack of evidence. He considered the controversy regarding the identity with modern forms of the mummified animals brought back from Napoleon's expedition to Egypt. Weighing up both Cuvier's argument against transformism based on their identity with modern species and Lamarck's reposte that insufficient time had gone by since the time of the pharaohs for any change to be observable, he concluded that no definitive answer was possible without more evidence. Summing up, he concluded that: 'We cannot, in short, subscribe to the full extent of this author's doctrine; because the facts and analogies, with which we are acquainted, would not bear us out in the support of it'.[53] In short, while Lamarck's theory was plausible and compelling, there was as yet insufficient evidence to decide the question.

In a second part of the review, not published until almost two years later, Muirhead gave a similar verdict on the question of spontaneous generation that was so central to Lamarck's theory. He concluded:

We are at all events, compelled to admit that the origin of various living beings, as of some of the byssi and mucors, infusorial animals, hydatids, intestinal worms, &c., is veiled in the most profound mystery, and seems to be scarcely reconcilable to our common notions of vegetable and animal reproduction.[54]

Muirhead's review demonstrates that right from the start reactions to Lamarck's works in Scotland were quite nuanced. Far from rejecting Lamarck's transformist theories out of hand, Muirhead was prepared to consider them as a plausible explanation of the diversity of living beings. Crucially, he also saw no conflict between them and his religious principles. Only the lack of compelling evidence stood in the way of his accepting Lamarck's conclusions.

Although he would have been aware of older, eighteenth-century varieties of transformism, which he would have learned about in John Walker's lectures, there is no evidence that Jameson knew about the work of Lamarck until around 1813.[55] Jameson's preface to the first English edition of Cuvier's *Discours sur les révolutions de la surface du globe*, published in that year under the title of *Essay on the Theory of the Earth*, contained an explicit reference to Lamarck. Here, Jameson stated:

> Some naturalists, as La Mark [sic], having maintained that the present existing races of quadrupeds are mere modifications or varieties of these ancient races which we now find in a fossil state, modifications which may have been produced by change of climate, and other local circumstances, and since brought to the present great difference by the operation of similar causes during a long succession of ages, – Cuvier shews that the difference between the fossil species and those which now exist, is bounded by certain limits; that these limits are a great deal more extensive than those which now distinguish the varieties of the same species; and, consequently, that the extinct species of quadrupeds are not varieties of the presently existing ones.[56]

It seems likely that Jameson's acquaintance with Lamarck's thought at this time was relatively superficial and probably at second hand. There are three pieces of evidence for this. The first is his doubly incorrect spelling of Lamarck's name, which would seem unlikely if Jameson had been well acquainted with his works. He spelled Lamarck's name as it is pronounced rather than it is written, so he may only have heard his ideas discussed rather than read about them. The second is his apparently limited knowledge of Lamarck's theory. The quotation above argues that fossil forms cannot be the ancestors of existing animals because they are too different to be varieties of the modern species. In doing this Jameson credited Lamarck only with the idea that fossil animals are 'mere modifications or varieties of these ancient races'. In consequence, the argument presented by Jameson turned solely on whether the modern forms could or could not be varieties of the fossil animals. As we have noted above in relation to John Walker's lectures, the production of new varieties through the influence of physical conditions was relatively uncontroversial in this period. For Jameson, the 'present great difference' between living and fossil forms clinched the argument that these were not just varieties of the same species. The possibility that one species could have transmuted into another completely distinct species was not even addressed. His argument curiously neglects Lamarck's belief that transmutation was possible between species as well as between varieties of the same species, and indeed that this was central to his theory: a curious omission if Jameson was thoroughly familiar with Lamarck's work at this time. It would be easy to conclude on the basis of this passage that Jameson had entirely missed the true import of Lamarck's theory. It is, of course, conceivable that Jameson was simply making rather loose use of the term 'variety', but this seems unlikely given the clear distinction made by Walker. The third reason for concluding that Jameson knew little of Lamarck's theory at this time is his apparent belief that the principal driver for transmutation in it was change in the physical conditions of life, an impression that even a passing acquaintance with Lamarck's writings would have dispelled.

We know from the reference above that Jameson had heard something of Lamarck's work from as early as 1813. Among Jameson's papers, there are to be found some clues as to how and when Jameson may have first come into direct contact with Lamarck's transformist theories. Two receipts survive there for volumes of Lamarck's *Histoire naturelle des animaux sans vertèbres* from Treuttel & Co. of London, dated 25 May 1822 and 11 April 1823.[57] We therefore know that Jameson certainly owned a copy of this work, which devoted almost its entire first volume to a detailed exposition of Lamarck's theory of the transmutation of species. According to remarks by George Johnston on John Fleming's *History of British Animals* (1828), published in the *Edinburgh New Philosophical Journal* in 1828, Lamarck's *Histoire naturelle* was 'in general use among naturalists in this country; and it is necessary that the student should be acquainted with its language or synonymes, whether he may choose to adopt them or not'.[58] Jameson was therefore not alone in appreciating the value of this work. We know that the Plinian Society also had a copy of the *Histoire naturelle* in its library from 1827.[59] Remarkably, a handwritten translation of parts of it into English in the handwriting of Jameson's assistant at this period, William MacGillivray, also survives among Jameson's papers.[60] The paper this is written on is watermarked 1821, so it probably does not long postdate the purchase of the book. It is certainly not likely that it was written after August 1831, when MacGillivray succeeded Robert Knox as conservator of the museum of the Royal College of Surgeons of Edinburgh.[61] The translation was made from the first French edition (1815–22), as a page reference to a 'table of articulated and inarticulated animals' on page 457 in Chapter XII (f.5) refers to the pagination of that edition, which differs from that of the second edition. We can safely conclude that Jameson bought his copy soon after the publication of the final volume. The first folio begins mid-sentence with material from page 400 of volume 1, from which it is evident that part of the original manuscript has been lost. It is not possible to determine whether the entire book was ever translated, as each chapter has been paginated separately in the copy. What is apparent from the pagination, however, is that there are a number of lacunae in the manuscript. From the surviving sections, we can, however, say that it included at least parts of volume 1 to volume 6, part 1. Despite its fragmentary nature, the existence of this translation is conclusive evidence that Jameson was paying attention to the work of Lamarck in the early 1820s.

The single most important commentary on Lamarck's theory published in Edinburgh in the early decades of the nineteenth century is the anonymous 'Observations on the nature and importance of geology', published in 1826 in the *Edinburgh New Philosophical Journal*. This paper is significant in that, as Secord has pointed out, it is is the 'earliest favourable reaction to Lamarck in a British scientific periodical', if we set aside the somewhat equivocal review of the *Philosophie zoologique* by Muirhead.[62] What did this important paper have to say about the theories of Lamarck? First, the author observed that, in the past, naturalists had attempted 'to arrange the species of animals, sometimes according to a scale of gradation, and sometimes according to a reticulated form, without giving any distinct account of the meaning of such an arrangement'.[63] The question, then, was to give meaning to the relationships that had been observed between the different species of animals. He then introduced the theory of Lamarck, 'one of the most sagacious naturalists of our day', as a possible answer to this problem.[64] Then followed a brief account of Lamarck's views on the spontaneous generation of infusory animals and simple worms, and the evolution of all existing animals from these first primitive forms under the influence of external circumstances.[65] Later in the article the author went on to introduce the evidence of domestic animals and cultivated plants in support of Lamarck's theory and against the supposed immutability of species, as had Muirhead in his review of the *Philosophie zoologique* fifteen years earlier.[66] He then sounded a note of caution, before concluding that Lamarck's ideas offered the best explanation of the relations between the species of animals currently available:

> Although it should not be forgotten, that this meritorious philosopher, more in conformity with his own hypothesis than is permitted in the province of physical science, has resigned himself to the influence of imagination, and attempted explanations, which, from the present state of our knowledge, we are incapable of giving, we nevertheless feel ourselves drawn towards it, as these notions of the progressive formation of the organic world, must be found more worthy of its first Great Author than the limited conceptions that we commonly entertain.[67]

The paper is unusual for transformist writings from the period in that it directly and explicitly engaged with Lamarck's theories, while the others generally avoid naming their sources. The anonymous author made a strong case that geology alone could give concrete evidence for the truth of Lamarck's theory, and noted that the gradual appearance of increasingly perfect forms over geological time already provided strong support for the transmutation of species. He then tackled the subject of extinction, first asking if changes in physical conditions might have led to the extinction of some species, whether 'from a change in the media in which organic creatures lived, and from powerful causes operating upon them, their power of propagation may be weakened, and at length become perfectly extinct?'[68] This is counter to the view of Lamarck, who considered that no species ever became extinct, except perhaps for those occasionally driven to extinction by human agency. Instead, according to Lamarck, apparently lost fossil species had either transformed into new ones over time, or had simply not yet been discovered in living form.[69] No firm conclusion was reached by the author on the subject of extinction, however, and he was also prepared to countenance Lamarck's idea that 'many fossil species to which no originals can be found, may not be extinct, but have gradually passed into others'.[70] Finally, and before moving on to discuss the utility of geology for solving the problems raised by the distribution of species around the globe, he addressed the observation made by Cuvier that the mummified ibises brought back by Geoffroy in the aftermath of Napoleon's ill-fated expedition to Egypt were identical to the modern species.[71] The author dismissed this objection, utilising the same argument we have already seen used elsewhere by Jameson, Geoffroy and Lamarck, that the time that had passed since the time of the pharaohs was insufficient for any change to be evident: 'what are a few thousand years to which the mummy refers, in comparison with the age of the world, as its history is related by geology'.[72] What emerges is a broadly Lamarckian model of transformism, although the role played by change in physical conditions, probably showing the influence of the Wernerian model of earth history, is more reminiscent of Geoffroy's theories.

Unlike Muirhead, the author of the 'Obervations' was convinced he had the evidence required to validate Lamarck's theory. This evidence came from the fossil record, for '[t]he doctrine of petrifactions, even its present, imperfect condition, furnishes us with accounts that seem in favour Mr Lamarck's hypothesis'.[73] The successive appearance of different forms of life in the geological record in ascending order of perfection provided strong evidence for Lamarck's transformist theory. Thus, for the author of the 'Observations', geology provided the proof that Muirhead had felt was lacking.

Outside the University, there is some evidence that Lamarck's work was also known at the Royal College of Surgeons of Edinburgh. The work of L. S. Jacyna has revealed that Lamarck's theories were discussed among a group of medical practitioners identified by him as 'philosophic Whigs', who were active in Edinburgh in the early decades of the nineteenth century. Jacyna has shown that John Thomson (1765–1846), professor of surgery at the Royal College of Surgeons from 1804 to 1821, regius professor of military surgery at the University from 1806 and professor of general pathology from 1832 to 1841, presented a 'carefully edited version of Lamarckism' in his lectures.[74] However, despite 'his numerous references to Lamarck . . . he never hinted at the possibility of transformism'.[75]

Among the lecturers who taught in Edinburgh's extra-mural medical schools in the 1820s and 1830s, Robert Grant stands out as the best known disciple of Lamarck. Despite not making explicit reference to Lamarck in his published work in this period, we know that Grant was sympathetic to Lamarck's transformist theories, not only from his own open espousal of transformism but also from the testimony of Charles Darwin. In his autobiography Darwin, who got to know Grant well in his time at the University of Edinburgh, recalled how they used to go on invertebrate-collecting trips together on the Firth of Forth. He recounted how one day while they were on such a collecting trip, probably in late 1826 or early 1827, Grant 'burst forth in admiration of Lamarck and his views on evolution'.[76] Despite this, there seems to have been nothing specifically Lamarckian about Grant's own transformist speculations, which are closer to the theories of Geoffroy.

A paper by Grant published in David Brewster's *Edinburgh Journal of Science* in 1828 on the reproduction of the sponge *Lobularia digitata* provides some fascinating insights into Grant's views on the origin of life and the fundamental constitution of living things. It also sheds light on the relationship between his transformist ideas and those of Lamarck. Grant first noted that the ova of the sponge were transformed from 'moving, irritable, and free condition of animalcules, to that of fixed and almost inert zoophytes'.[77] He went on to observe that freshwater algae (Confervae) had been seen to resolve themselves into animalcules and that these animalcules could then reunite to reconstitute the plants. He then expressed the belief that 'Mosses and Equiseta are found to originate from confervae ... and all the land confervae with radicles appear to pass into the state of more perfect plants'.[78] The implication is that there is a gradual development within the plant kingdom, and by extension in the animal kingdom too, from animalcules through simple forms to the most perfect types. The animalcules therefore represent the starting point for the development of higher forms and the basic units of life from which higher forms are constructed.

P. R. Sloane has taken this paper, together with evidence from Grant's introductory lecture at the University of London that the 'Animal and Vegetable Kingdoms are so intimately blended at their origins, that Naturalists are at present divided in opinion as to the kingdom to which many well-known substances belong', to suggest that Grant postulated a common origin for plants and animals.[79] If this were the case it would put him entirely at odds with Lamarck, who wrote of animals and plants in his *Histoire naturelle des animaux sans vertèbres* that 'the two branches of which I have just spoken are in reality separated from each other at their base, and a positive characteristic that relates to the chemical constitution of the bodies on which nature has worked makes an eminent distinction between the beings which are embraced by one of its branches and those which belong to the other'.[80] Sloan has suggested that Grant's ideas were derived from Friedrich Tiedemann, who wrote that '[o]ne might even be almost tempted to believe that, in certain circumstances, the most simple vegetable and animal forms may pass from one to the other. Confervae are resolved into infusoria, and infusoria produce confervae by their union'.[81] Sloan has argued that Grant's use of the term 'zoophyte' to designate sponges, a term abandoned by Lamarck because he rejected the idea of intermediate forms between plant and animals, indicates that Grant saw sponges as intermediate plant-animals. However, Grant consistently referred to sponges as 'animals' in his papers on them in the *Edinburgh Philosophical* and the *Edinburgh New Philosophical Journals* in the 1820s and nowhere suggested that they should be treated as intermediate forms between the two kingdoms. It seems more likely that Grant considered the link between plants and animals to exist at the level of primordial animalcules rather than more complex organisms. His use of the term 'zoophyte' may simply have been a matter of convention. What is undeniable is that Grant, like Lamarck, saw animalcules as the origin of all animal life on earth. As he was to say in his lectures as professor of comparative anatomy at University College London, which were subsequently published in *The Lancet* in 1833–4: 'When we speak of animals low in the scale, it is equivalent to our speaking of animal forms that have existed in the primitive conditions of this planet; for everything shows, that this kingdom itself has had a development from the most simple forms.'[82]

Grant's last signed article in any of the Edinburgh journals was a piece in the *Edinburgh New Philosophical Journal* in 1834, and his final contribution to the *Memoirs of the Wernerian Society* was in 1832, so it seems that his association with Edinburgh natural history circles may have gradually faded into the background in the course of his first decade in London. Nevertheless, his views did not seem to have radically changed on leaving Edinburgh. In his introductory lecture in 1828 he asserted 'that animal life originated and was developed in the bosom of the deep'.[83] He went on to note: 'From numerous experiments, Naturalists have been led to believe that the simplest organized bodies, as *Monads* and *Globulinae*, originate spontaneously from matter in a fluid state, and that these simple bodies, of spontaneous origin, are the same with the gelatinous globules which compose the soft parts of Animals and Plants.'[84] As in his 1828 article for Brewster's journal, Grant is in accord with Lamarck on the importance of spontaneous generation, but parts company with him on the fundamental distinctness of animals and plants.

Grant also differed with Lamarck on the subject of extinction. He clearly believed that many species had become extinct over the course of geological time, or as he puts it 'Numberless species, and even entire *genera* and tribes of animals, the links which once connected the existing races, have long since begun and finished their career.'[85] By contrast, Lamarck believed that the species found in fossil form had either been transformed into the ones we now find inhabiting the earth, or had simply yet to be discovered in living form. Grant came very close to the catastrophism of Cuvier when he remarks that '[b]y thus pointing out the extensive and terrible catastrophes to which the Animal Kingdom has often been subjected, we are enabled to perceive a cause of the many apparent interruptions in the chain of existing species'.[86] Once again, this statement is quite at odds with the gradualism apparent in Lamarck's vision of the history of life. It is clear that his transformist view were strongly influenced by Lamarck, although he does differ from him in a number of crucial regards.

Not all followers of the new philosophical anatomy in the extra-mural schools were as enthusiastic about Lamarck's theories as Darwin implies Grant was. According to Philip F. Rehbock, Robert Knox 'did not feel comfortable with Lamarck's or Geoffroy's views of evolutionary descent'.[87] This may be true as far as it goes, but Evelleen Richards has demonstrated that, while Knox might not have subscribed to Lamarckian transformism, he seems to have developed his own rather eccentric theory of organic descent.[88] His belief in the transmutation of species and his rejection of Lamarckian transformism were both acknowledged by Baden Powell in the mid-1850s when he wrote that Knox, 'one of the most zealous supporters of the principle of transmutation in this country, speaks very slightingly of Lamarck'.[89] And indeed he did; in an article published in *The Lancet* in 1855 Knox gave the opinion that '[t]he wild conjectures of Le Methrie [Julien Offray de la Mettrie (1709–51), the materialist author of *L'Homme Machine* (1748)] and Lamarck were written in a style of romance, excluding them from the sober field of science'.[90]

The hostility of Knox to the theories of Lamarck still, however, needs to be explained. There are two likely reasons for his response. First, Knox was profoundly hostile to the notion of progress in the natural world, and the place of man at the pinnacle of creation. He looked forward to a time when the 'boast about the higher characters of the present organic races will be abandoned, and the law of development and progress simply stated as it is, without a reference to successive *improvement*; for *successive improvement* implies a final purpose'.[91] Later in the same book he went on to write, in his usual acerbic manner, that the 'world, for countless thousands of years, was inhabited only by fishes; could they have spoken, and left us records, we should have found, no doubt, that they considered themselves as the most perfect of all Nature's works, and the beings for whom the seas, at least, if not the dry land, had been made'.[92] It seems that Lamarck's rather triumphalist account of the ascent of man may have sat rather ill with Knox's distrust of theories of progressive development.

Second, Knox disagreed with Lamarck's use of the principle of the inheritance of acquired characteristics in his theory, the possibility of which Knox expressly denied. This denial that over time species could be modified by their conditions of life was linked to his belief in the unshakable permanence of human races, which came to preoccupy him more and more in later decades. In *Great Artists and Great Anatomists* he dismisses Lamarck's mechanism for the transmutation of species in the following words: 'Lamarck's idea was that organization was the result of function, and not function the necessary result of form; that an animal was aquatic, not by the nature of its organs but became so, acquiring a fitting organization by its being forced to live in water. This view was wholly theoretical and met with no respect.'[93] Knox's choice of words here is intriguing. To say that Lamarck's ideas 'met with no respect' is by no means the same as saying they were unworthy of respect. As is often the case with Knox's writings, his true opinions are often veiled behind a highly rhetorical, deliberately provocative style, with frequent apparent inconsistencies and contradictions. Having said that, it is clear that Lamarck's particular brand of transformism did not meet with his approval. However, Knox's dismissal of Lamarckian transformism should not be taken to imply that he was a confirmed opponent of transformism *per se*.

Interest in Lamarck's theories among Edinburgh's extra-mural lecturers was not confined to Grant and Knox. John Fletcher also wrote a detailed critique of Lamarck's theories in his *Rudiments of Physiology* (1835–7). The picture painted by Fletcher is the skewed one found in the writings of many of Lamarck's critics, which may ultimately be traced back to Julien-Joseph Virey's influential *Nouveau dictionnaire d'histoire naturelle* (second edition 1816–19) and Cuvier's mean-spirited eulogy to Lamarck, published in the *Edinburgh New Philosophical Journal* in 1836. Corsi has established that other British natural historians had been misled by Virey's account of Lamarck's theory.[94] Fletcher wrote that Lamarck 'traces all tribes of animals to the lowest zoophyte, and ascribes all the differences which they now display entirely to the different instincts which they have experienced, and the different efforts which they have severally made to gratify them'.[95] In common with many other critics of Lamarck, Fletcher's critique relied on an incomplete and inaccurate account of his theories. He must therefore be classed among those critics of Lamarck who based their rejection of his work on his supposed belief that living things will themselves to evolve, a crude caricature of his actual theories.

To cite another example of the superficiality of Fletcher's reading of Lamarck, he classed him as among 'those who conceive that every form of organized matter consists of a congeries of monads or organic molecules of precisely the same nature, and competent therefore to enter into the composition of any organized being'.[96] According to Fletcher, Lamarck shared this view with Turberville Needham, Buffon, Gottfried Reinhold Treviranus and Friedrich Tiedemann, among others. Of these figures, perhaps the one whose ideas are most closely represented by Fletcher's generalisation is Buffon, but the concept of identical 'organic molecules' plays no part in the theories of Lamarck. Although Fletcher's own religious affiliations are unknown, like many evangelical critics of Lamarck, his rejection of Lamarck seems to have been determined to some extent by his religious convictions, as is made clear from this following passage from one of the footnotes to the *Rudiments of Physiology*:

> The study of nature is the study of God's nature; and it is only they who have stopped on the threshold of this study, and have let in only light enough to render darkness visible, or who are evidently wrong-headed – and such men have existed from Epicurus to Lamarck – that have indulged in those flippant and irreverent remarks, the object of which is to shake our faith in truths which it must distress us to doubt, and wither us to disbelieve.[97]

However, Fletcher's religious arguments are rather different from those typical of evangelical critics of Lamarck and do not bear the distinctive stamp of evangelical theological preoccupations, such as their distrust of the idea of progressive development in a fallen world, or their outrage at any implied questioning of God's power to miraculously create new forms of life *de novo*. They appear rather to be of a more moderate, natural-theological character. All of the Edinburgh figures whose reactions to Lamarck's theories I have discussed so far have one thing in common: as far as their religious views are known, they all seem to have been moderate in their religious views, if not theists. As we have seen, these individuals generally seem to have been rather open to Lamarck's theories when not actively accepting of them. The reception he received from evangelicals was often rather different. During the first half of the nineteenth century, Edinburgh was home to an extraordinary group of individuals belonging to the Evangelical Party of the Church of Scotland who took a strong interest in science and natural history.[98] These included such figures as the minister and natural historian John Fleming, the natural philosopher David Brewster, the minister and reformer Thomas

Chalmers and the geologist and journalist Hugh Miller. Theological considerations had a strong influence on their reactions to the theories of Lamarck, as they did on others who shared their evangelical perspective.

Evangelical theology was seen as being incompatible with transformism for two main reasons. First, evangelicals generally viewed the world as fundamentally corrupted and depraved as a result of the Fall. Many evangelicals therefore came to deny any possibility of progress, whether in the natural world or in human society, other than personal spiritual progress, or the amelioration of society that they believed resulted from it.[99] Second, transformism seemed to suggest that God was content to rely on secondary causes in his governance of the world and appeared to deny him the power to also intervene in the natural world by supernatural means when he chose to create new species. While this was a matter of indifference to moderate figures such as Muirhead, it was anathema to evangelicals, as it conflicted with the evangelical emphasis on the absolute power of God. Although these were not the only theological arguments against transformism, they were the ones most generally emphasised by evangelical critics. In the remainder of this section I will be examining the reactions of two evangelical critics to the transformist theories of Lamarck during the 1820s. In this period evangelical critiques were often relatively nuanced and tended to concentrate on the scientific weaknesses of the theory as much as the theological problems that they raised. However, from the early 1830s onwards there was a definite hardening of evangelical attitudes; these later developments will be discussed in more detail in Chapter 6.

An early Evangelical response to Lamarck comes from the *Memoirs of the Wernerian Society* and constitutes the only explicit reference to Lamarck to be found there. This was contained in a paper by the evangelical minister James Grierson (1791–1875), which was given to the Society in February 1824. Here Grierson made reference to '[t]he original or infinitely small monadic animals and vegetables, Lamarck, and others, who hold the same system, tell us, gradually acquired different habits, became larger and more diverse from one another; and hence all the animals and vegetables we have now'.[100] He then goes on to dismiss Lamarck's ideas, 'which, if they do not evince much power of observation, or great accuracy of deduction, certainly shew no deficiency in power of fancy'. While Grierson was clearly no follower of Lamarck, his remarks at least demonstrate that his theories were known and discussed by members of the Wernerian Society. Lamarck's doctrines were dismissed by Grierson principally on the grounds of being fanciful speculation and therefore bad science, with no indication that any political or religious anxieties were aroused by the implications of his theories.

John Fleming has left us considerable evidence in print of his thoughts on Lamarck's theories. Fleming was very much part of Jameson's circle in the early decades of the century. Along with Jameson, Fleming was a founder member of the Wernerian Natural History Society. Jameson also joined with John Playfair and David Brewster in successfully proposing Fleming as a fellow of the Royal Society of Edinburgh in 1814. In the mid-1820s Fleming became involved in a lively polemic with William Buckland, reader in geology at the University of Oxford, over the nature of the Deluge.[101] He was the leading natural historian of the Evangelical party of the Church of Scotland, and as such after the Disruption he was the natural choice for the chair of natural history at the Free Church College in 1845. As Pietro Corsi has shown, Fleming shows himself to be surprisingly sympathetic to Lamarck's theories in his writings, while not necessarily sharing his conclusions.[102] In April 1820 a review of Lamarck's newly published *Histoire naturelle des animaux sans vertèbres* by Fleming appeared in the *Edinburgh Review*. Fleming started by praising Lamarck, recognising that '[h]is writings, which are now voluminous, are generally characterised by the research and ingenuity of his speculations, and by the clear and perspicuous language in which he has embodied them'. But he added that 'they also betray a decided propensity to generalize on assumed or deceptive premises, and they are all, more or less, tinctured with the influence of a few leading and favourite doctrines, which seldom rest on very stable foundations'.[103] It soon emerges that the key doctrine that Fleming objected to was Lamarck's materialism. In particular, jumping to the fourth section of the introduction to Lamarck's book, he took issue with his materialist explanation of mental phenomena to be found there. He criticised Lamarck for his supposed advocacy of the principle that the faculties of the mind are localised in particular areas of the brain, noting that 'we find him confidently asserting the doctrine, that every mental faculty has its appropriate organ, without which it cannot exist: but the nature of the union of matter and spirit in our own constitution is too mysterious to enable us implicitly to adopt any such proposition.'[104] It soon becomes apparent that Fleming's profound hostility to these ideas was rooted principally in his religious convictions. According to Fleming, Lamarck's ideas were 'subversive of those sublime and consoling views of religion which teach us, that mind may exist and act independently of matter altogether, and may be combined with it, or detached from it, at the will of the Sovereign ruler'.[105] As evidence that the mind is not a product of the material structure of the human brain, Fleming notes that 'the intellect, for example, has been observed to continue unimpaired, when a large portion of the brain has been obliterated or removed'.[106]

At this point Fleming seemed to be declining to turn from Lamarck's theory of the mind to address the transformist theories that Lamarck builds on the foundations of his materialist premises. Instead he stated that, 'impressed as we are by the conviction, that the author's premises are often extravagant, or erroneous, we are the less solicitous to accompany him, step by step, in this preliminary dissertation'.[107] But then, despite his avowed hesitancy, he did in fact go on to launch the following concerted assault on Lamarck's transformist theory:

> Even thought and imagination are represented as mere physical appearances; and new organs are formed, by mechanical means, *in consequence of a strong feeling of their need*, of the performance of the functions to which they are destined. But does a physical feeling create a physical organ? or, if this strong feeling is not physical, then are not *all* the phenomena of mind physical? Be it, however, what it may, when did the most intense wish, or feeling, of any human being, generate an additional organ to his original frame? or when did the most ardent desire of an unfortunate culprit to fly from the pursuit of justice, furnish him with wings? Farther, the existence of an ascending scale, from the more simple to the more complex animal structures does not necessarily imply, that the different tribes of living creatures were successively produced in that order, or that nature was compelled to limit her efforts to the scanty and imperfect, before she could progressively advance to the more ample and finished forms.[108]

It is interesting to note that, like Fletcher, Fleming misrepresented Lamarck's theory as implying that animals in some sense willed themselves to evolve. This is identical to the argument that Cuvier was to use to damn Lamarck's ideas in his notorious eulogy to him in 1832.[109] As Lamarck was still alive in 1820, Cuvier cannot have been his source, and Fleming may instead have picked up this interpretation from the second edition of Virey's *Nouveau dictionnaire d'histoire naturelle*. Fleming's criticism of Lamarck's theories in 1820 was not based principally on religious principles, or its lack of conformity with the Mosaic account of creation. Rather it was rooted in a sustained critique from a scientific perspective of Lamarck's mechanism for the transmutation of species and his vision of the history of life, based on Fleming's own rather skewed interpretation of his theories. When he did attack Lamarck's theory as incompatible with true religion earlier in his review it is clear that he was criticising Lamarck's materialistic theory of the mind, not his transformist views. His main concern there was to defend the existence of an immaterial vital principle against Lamarck's materialism on religious grounds, a preoccupation he shared with John Barclay. Finally, it is important to note that in this review

Fleming's appraisal of Lamarck's work was far from being altogether negative. He does not seem to consider the ideas contained in it as dangerous, or counsel his audience against reading the book. On the contrary, he concludes that this is 'a work, which, for all its defects, promises to hold a distinguished station in the library of the zoologist, and to impart both an impulse and facility to the study of the various tribes of beings of which it treats'.[110]

Two years after the publication of his review of the *Histoire naturelle*, Fleming published his *Philosophy of Zoology* (1822), the title of which strangely echoes that of Lamarck's *Philosophie zoologique* of 1809. Once again, we find Fleming concerned to defend the existence of an immaterial vital principle against Lamarck's conception of life as a consequence of the organisation of matter. For Fleming as for Barclay, the vital principle consisted of an immaterial essence that gave life to inanimate matter, a view fundamentally opposed to Lamarck's theory, which made life a result of the action of the same laws of nature that applied to non-living matter. In Chapter 2 of his book, 'On the peculiar characters of organized bodies', Fleming remarked that in his *Histoire naturelle des animaux sans vertèbres* Lamarck 'refers some of the movements which are here considered as indicating the existence of irritability in plants, to the influence of the mechanical or chemical powers'; however, in Fleming's opinion, 'these different actions ... occur in connection with the vital principle, and their entire dependence on the laws of inorganic matter is a gratuitous assumption'.[111] In opposition to Lamarck's materialist vision of life, Fleming then went on to expound his own vitalist theory of the 'Living or Vital Principle', which, 'so far as appears to our senses, can only reside in organized bodies. The connection is temporary, and may be dissolved by various circumstances and it is capable of being divided or multiplied by the process of generation'.[112] It is responsible for the generation and development of the organism and 'in the formation of an organized body, acts in direct opposition to the laws of chemistry or mechanics'.[113] For Fleming the vital principle seemed to correspond to the essence of a species in an Aristotelian sense; as a consequence there were 'as many different kinds of vital principles, as there are species in nature'.[114]

In 1829 Fleming again addressed the theories of Lamarck in a review of J. E. Bicheno's *Systems and Methods in Natural History* (1827), published in the *Quarterly Review*. By the time he wrote this review, Fleming's views seem to have subtly hardened against transformism. After giving a lengthy account of Lamarck's theories, Fleming argued that Lamarck introduced needless complication to the story of life, asking why God could not have created 'Man directly, as easily as a Monas'.[115] How God chose to operate in the natural world was clearly not the object of indifference that it had been for Muirhead. Given God's infinite power, he would surely have brought his creation into being directly rather than relying on secondary causes. While Fleming acknowledged that Lamarck had 'succeeded in making some converts', including several respected naturalists whom he did not name, he ultimately concluded that 'the whole scheme, as an exposition of the plan of procedure, is so obviously a dream of the imagination, that one may well be surprised to find it occupying a place in the records of science'.[116]

To sum up Fleming's critique of Lamarck, it must first be noted that in the early 1820s he was prepared to engage constructively with transformist theories rather than condemn them out of hand, especially in his *Philosophy of Zoology*. He not only seemed prepared to consider the possibility that there was some merit in transformist ideas, but conceded that they had evidence in their favour. It is also worthy of comment that his criticisms of transformism itself, as opposed to materialism, were based not on religious, moral or political criteria, but on the grounds of scientific plausibility and a lack of fit with empirical evidence. Although he came down against the transformist theories of Lamarck, he clearly had immense respect for his work as a zoologist and was very far from condemning him as a dangerous radical. This relatively tolerant attitude seems to have given way to a more inflexible attitude by the end of the 1820s. Fleming's moderate stance on Lamarckism during the 1820s is entirely consonant with his friendship with Robert Grant, the most openly transformist of Edinburgh natural historians. Later critiques of Lamarck from evangelical sources in the 1830s and 1840s would become increasingly hostile and put more emphasis on strictly theological objections.

## THE IMPACT OF GEOFFROY'S THEORIES IN EDINBURGH

The initial impact of Geoffroy's theories seems to have been limited to a specialist audience of comparative anatomists who were influenced by his brand of transcendental anatomy. These were principally teachers in Edinburgh's extra-mural medical schools, some of whom had studied with Geoffroy in Paris. These extra-mural lecturers were then in a position to transmit these ideas to the hundreds of medical students that passed through their classes every year. We know from the records of some of the thriving student societies that existed in Edinburgh in the 1820s that these exciting new ideas went on to be vigorously discussed and debated among medical students in their meetings.

In 1829 Jameson published an anonymous report in the *Edinburgh*

*New Philosophical Journal* of a memoir read by Geoffroy before the French Academy of Sciences and published in the *Mémoires du Muséum d'Histoire Naturelle* the previous year.[117] This report was attributed to Grant by Desmond in a 1984 paper.[118] However, Pietro Corsi has recently demonstrated beyond doubt that the paper is in fact a direct translation of an anonymous article that appeared in the French newspaper *Le Globe*.[119] The article gives a detailed account of Geoffroy's transformist theories and supports his belief that changes in the composition of the atmosphere drove the transmutation of species. The content of this paper was clearly of great interest to Jameson, as on 25 April 1829 he 'gave an account of the doctrines of Geoffroy Saint-Hilaire on the analogy between extinct animals and those now living' to the Wernerian Society, although sadly no record of exactly what Jameson had to say about Geoffroy's ideas has survived to enlighten us as to the opinions he expressed on that occasion.[120] However, from the brief description of the talk from the minutes quoted above it seems almost certain that his paper focused principally on Geoffroy's transformist theories, which would surely have been congenial to Jameson, based on what we know about his own views. Given the coincidence of dates between Jameson's paper to the Wernerian Society and the publication of 'Of the continuity of the animal kingdom' in his journal, which appeared in the April–June 1829 number, it seems highly probable that the paper he gave was largely based on that article.

While Jameson was clearly intrigued by Geoffroy's transformist speculations, it was in the extra-mural medical schools that Geoffroy's theories had their greatest impact. Robert Grant and Robert Knox, who both taught at John Barclay's anatomy school, had studied in Paris with Geoffroy, and visited the city repeatedly in subsequent years to renew old acquaintances and catch up on the latest developments in comparative anatomy. In a paper published in 1826 'On the structure and nature of the Spongilla friabilis' in the *Edinburgh Philosophical Journal*, in which Grant makes his transformist views explicit for the first time in print, we find clear evidence of the influence of Geoffroy's ideas on Grant. Towards the end of the article, he speculated regarding the relationship between the freshwater sponge *Spongilla* and the more complex marine sponges. While marine sponges had changed to adapt to the changing composition of sea water, which had become increasingly saline over time, freshwater sponge, 'living constantly in the same unaltered medium, has retained its primitive simplicity'.[121] Grant here gives a concrete example of the principle expounded by Geoffroy in his 'Organisation des gavials' that when the 'physical and chemical agents' to which an organism is exposed remain the same, so does the development of the organism, but when conditions change, the development of the organism exposed to these new conditions will be modified by them, provided the change is not so great as to kill it.[122] We know that Grant was an enthusiastic disciple of Geoffroy's views on unity of form in comparative anatomy and had got to know him well during his trips to Paris in the 1820s.[123] Unlike Lamarck, whose views on geology were essentially uniformitarian, Geoffroy believed that there had been a gradual but profound change in the composition of the atmosphere of the earth over geological time, and that this was the motor for the transmutation of species.[124]

References to Geoffroy's work continue after Grant's move to London in 1827. He made explicit note of Serres' and Geoffroy's recapitulation theory early in his *Introductory Lecture* to his course at the new University College London, but without linking it to an evolutionary succession. Instead, he simply noted that 'by comparing the human brain in the earliest stages of development, with the permanent forms of that organ in quadrupeds, birds, reptiles, and fishes, the most singular resemblances have been discovered, which throw new light on the gradual development of that organ in the most perfect animals, and on its remarkable structure in the inferior classes'.[125] This debt to Geoffroy is even more apparent elsewhere in an anonymous set of student's lecture notes from the session 1833/4 that survive from Grant's years in London. Here he went further, in that he explicitly linked foetal development to the appearance of new forms over geological time:

> the development of every organ of the human body can be traced thro' all its successive stages in the great body of the animal kingdom & the form w[hi]ch an organ presents in each of the lower classes corresponds with its condition at some period of the human embryo. But the researches of the Com[parati]ve Anatomists are not confined to existing races of animals. In the remains of animals entombed in the ancient strata of the earth, he is enabled to trace phases of organic development on the surface of our planet, w[hi]ch have long preceded the existing forms, & have been distinct ant[erio]r to the existence of our race, & of all the vertebrate tribes.[126]

A very significant proportion of Grant's lectures on comparative anatomy at the University of London seem to have been taken up with an exposition and defence of Geoffroy's theory of unity of plan. In the notes from the 1833/4 session, he followed Geoffroy in finding 'a unity of plan in the organization of the whole animal kingdom'.[127] This implies a radical rejection of Cuvier's four *embranchements*, or fundamental divisions of animal life. He went to particular pains to establish that the cephalopods provided a link between the invertebrates and the vertebrates, providing an abundance of anatomical evidence in support of his claim; for example: 'As we find in the class of fishes, remains of the ext[erna]l shells in the form of calcareous scales or plates or solid spines, so we find in the cephalopods the soft cartilaginous rudiments of the vertebral column, which is met with in the myxene & lampreys & other of the lowest cartilaginous fishes.'[128] This very question was the bone of contention between Geoffroy and Cuvier in their famous debate at the Académie des Sciences in Paris in 1830. According to the printed syllabus for Grant's 1830 lecture course in 'Comparative Anatomy and Zoology', the zoology course ended with a section on 'Relations between the Extinct and Existing species of animals. Revolutions in the Animal Kingdom indicated by Fossil Animals'.[129] It would be interesting to know what the content of this final lecture was, but sadly no notes from this part of the course seem to have survived.

We have seen above that Robert Knox could at times be quite disparaging regarding the theories of Lamarck. However, a relatively early reference to transformism in Knox's writings that can give us a more nuanced view of his attitude towards transformism appeared in his translation of the lectures of De Blainville on comparative osteology, published in *The Lancet* in 1839. Although Lamarck is mentioned here, the focus is on the ideas of Geoffroy. In a note appended to these lectures, which merits quotation in its entirety, Knox stated:

> The lapse of time passing for nothing with modern geologists, Lamack [sic], Geoffroy (St. Hilaire), and others, took advantage of such views, by proposing a doctrine in which the specific distinctions of animals was set aside. The fossil Sawrians [Saurians, referring to extinct giant reptiles] became in their eyes, the progenitors of the recent crocodile [sic]; and although such theories were altogether *speculative*, and positively contradicted by all human experience and chronology, yet both being, as it were, but a drop in the great ocean of time, these distinguished naturalists conceived themselves to be at liberty to set aside all human experience and human records, and appealed to the obvious changes which have taken place in the surface of the globe, as indicative of an immeasurable antiquity, and of a sufficient lapse of time and change of circumstance to effect all manner of alterations in the primitive creation.[130]

The statement in the passage quoted above that the transformist theories of Lamarck and Geoffroy were 'altogether speculative, and positively contradicted by all human experience and chronology' was used by Rehbock to illustrate Knox's hostility to transformism.[131] Put back in its context, we can see that things are not quite so simple. What Knox in fact said was that, although transformism was 'contradicted by all human experience and chronology', the vast span of geological time makes the time frame of recorded history quite inadequate when discussing the history of life on earth. Despite the typically slightly mocking tone of this note, the only concrete criticism that is raised against transformism is that it is 'speculative', which hardly constitutes a conclusive refutation. This passage in fact is a good sample of Knox's rather elliptical rhetorical style, which often leaves his true opinion difficult to fathom.

What is clear from this passage is that Knox was familiar with Geoffroy's 1825 paper on his 'Recherches sur l'organisation des gavials', which suggested that modern crocodiles were the descendants of the fossil species found in France.[132] This paper was something of a landmark for Geoffroy for, as Toby Appel has remarked, before 1825 'Geoffroy had neither explicitly adopted special creationism nor mentioned evolution'.[133] Appel has concluded that there was no reason to believe that Geoffroy was a convinced transformist before 1825, although he had doubtless discussed the subject with his colleague Lamarck before then. It cannot therefore be assumed that Geoffroy's disciples, such as Knox and Grant, were aware of his transformist views before this date. Knox refers to this paper not only here, but repeatedly in some of his later works.[134] In his *Great Artists and Great Anatomists* (1852) Knox gave a quite differently nuanced account of his views on Geoffroy's theories from the one he had given in his translation of De Blainville's lectures:

> To Cuvier's theory of the 'Fixity of Species', as demonstrated by the drawings on the Egyptian tombs, Geoffroy objected, that 'as the surrounding circumstances had not changed, there existed no reason for a change in the Fauna'. He might have added, that the period referred to by Cuvier, in proof of his views, was but an instant in the duration of the globe.
>
> Convinced of the soundness of the basis on which Autenrieth, Goethe, and Geoffroy had constructed the great theories of Transcendental Anatomy, I hesitated not applying them constantly in all my researches in zoology, from 1820 inclusive: these principles were fully explained by me in three courses of lectures on Comparative Anatomy delivered to distinguished classes in 1825–26–27.[135]

Although, as often with Knox, the exact meaning of these two consecutive paragraphs when taken together is not altogether clear, two conclusions can safely be drawn from the arguments of each of them. From the first we know that Knox was aware that Cuvier's case against transformism based on historical evidence were fatally flawed. From the following paragraph we learn that Knox was a convinced disciple of Geoffroy during the 1820s. Knox nowhere explicitly said that he was a follower of Geoffroy's transformist theories, as opposed to his transcendental anatomy, but the juxtaposition of these two paragraphs is extremely suggestive, especially when later in the same book Knox stated explicitly 'I believe all animals to be descended from primitive forms of life.'[136] A few years later in an article in *The Lancet* he wrote: 'A last question remains – the origin of natural families: Have they been distinct from all time? I think not.'[137] These would appear to be unambiguous avowals that, at least in the early 1850s, Knox believed in some form of transformism. That, with Geoffroy, he was converted to a belief in the transmutation of species in the mid-1820s seems highly probable.

Although not a transformist, John Fletcher was an enthusiastic advocate of Geoffroy's transcendental anatomy among Edinburgh's extra-mural medical lecturers in the late 1820s and 1830s. For Fletcher, the 'classification of animals upon the principles, not of their general similarities of form and function, but of the fundamental structure of their organs, as proposed by Geoffroy Saint-Hilaire, De Blainville and others, is unquestionably much more profound and scientific than that of Cuvier'.[138] He went on to describe Geoffroy as the 'chief' of those who held that 'if not all plants, certainly all animals consist of the same number of organs, and these all fundamentally the same'.[139] Fletcher acknowledged that Geoffroy attributed the different types of animals 'to the greater or lesser degree of original development of a certain primitive type common to all'.[140] He went on to note that for Geoffroy this primitive type was not simply an ideal form, but had actually existed:

> Nor is the prototype on which each plant and animal, however elevated, is supposed to be based, conceived to be merely traceable in imagination through all the adventitious forms of the less perfect fabric; but to have actually existed, first preparatory to the general creation successively of the various tribes of organized beings, and again preparatory to the development of the embryo of each individual in its generation.[141]

This seems to be a rather garbled exposition of Geoffroy's transformist theories, although Fletcher correctly identifies the central role of foetal development in them. It is not clear from Fletcher's account that he had fully understood the details of Geoffroy's theory, or the nature of the mechanism for the transmutation of species that he had proposed. The theory he outlined does, however, appear to draw important elements from his theory, as well as from Buffon's concept of 'organic molecules'. Despite the detailed transformist theory he had presented, Fletcher finally rejected the idea, concluding:

> while it has nothing but the most vague and rambling presumptions in its favour, it is quite inconsistent with the generally immutable character of each tribe from the earliest periods of which we have any records; nor does the fact of many tribes being known to have formerly existed which have now perished

from the face of the earth, any more than that of many others now inheriting the earth which probably at one time had no existence, afford any proof of their mutual convertibility, or indicate any thing more than that the character of its inhabitants has, at different times, varied with that of the globe.[142]

Perhaps unsurprisingly, given the enthusiasm for his theories in the extramural medical schools, there is also evidence from the minutes of the Plinian Society to indicate that a significant number of the medical students at the University of Edinburgh were taking an interest in Geoffroy's transcendental anatomy in the early 1830s. On 15 February 1831 Thomas Shapter 'read a paper giving an analysis of Geoffroy St. Hilaire's Views of Unity of Organisation', after which Allen Thomson is reported to have 'read another paper on the same subject',[143] and on 25 April 'Mr French announced for next meeting a paper on Unity of Organization'.[144] Clearly Geoffroy's theories were both well-known and widely discussed in student circles in Edinburgh in the late 1820s and early 1830s. While the often scanty evidence of the minutes of the Society tells us nothing about the reception of Geoffroy's transformist theories by its members, as opposed to his ideas on unity of plan, the members must certainly have been aware of them.

One member of the Plinian Society who we know from other sources certainly took a positive interest in the theories of Geoffroy, was Henry H. Cheek. In 1830 a famous and acrimonious debate took place on the subject of unity of plan in the Académie des Sciences in Paris between Geoffroy, the great champion of the concept, and Georges Cuvier, its most famous opponent.[145] The confrontation between the two great comparative anatomists became something of a *cause célèbre* in intellectual circles across Europe, and was widely reported. Cheek's journal devoted considerable space to 'a controversy which has arisen between the two first zoologists of the age'.[146] The April number contained a detailed account of Cuvier's arguments against Geoffroy. In May followed Geoffroy's answer to Cuvier's charges. According to Toby Appel, 'even the supporters of Geoffroy agreed that Cuvier had had the upper hand in the debate' and that 'Cuvier could be said to have won the day'.[147] Although the articles in Cheek's journal evince a healthy respect for both of the rival anatomists, the reader is left in little doubt that the author favoured Geoffroy's arguments. Rather than acknowledge Cuvier the victor, the second article reports that 'M. St. Hilaire has consented to relinquish the discussion in the Academy; but, confident in the truth and novelty of his conclusions, he has determined to write a work, wherein he will controvert the opinions of M. Cuvier'.[148] In the event, Geoffroy's *Principles of Philosophical Zoology*, rushed into print in April 1830, did not definitively put an end to the dispute. Only the death of Cuvier in May 1832 finally prevented any further continuation of hostilities. There is little doubt where Cheek's sympathies lay, and there is much evidence for his espousal of both Geoffroy's principle of unity of plan and his transformist theories to be found throughout Cheek's journal.

In an article on the dugong in the December 1829 issue of the *Edinburgh Journal of Natural and Geographical Sciences*, Cheek appeared to combine a developmental vision of the history of life with the concept of unity of form associated with the philosophical anatomy of Geoffroy. In this article he noted that dugongs are 'nothing else than terrestrial mammalia, whose internal organs are concealed under the figure of a fish'. He went on to suggest:

> Speculation immediately suggests the geological fact, that fishes existed prior to the creation of the mammalia; and that the Omnipotent has passed by slow gradations from one series of organization to another; that the type or model on which all vertebrate animals are formed is essentially the same.[149]

Of course, that nature had brought forth species progressively over time in an ascending scale of perfection did not necessarily imply the transmutation of species, but was equally compatible with multiple creations. However, the fact that Cheek commented on the slow rate at which the series of organisation pass into each other strongly implies that it was a process of transmutation that he has in mind. It would have made little sense to talk about the speed at which each series had passed into the next if each had been an entirely separate creation.

Another reference to unity of type is found in the March 1831 issue in an anonymous editorial in Cheek's journal, almost certainly by Cheek himself, on the identity of the vascular arches of terrestrial vertebrates with the branchial arches of fish. This article appeared in a section of the journal entitled 'Zoological Collections'. Cheek noted that from the evidence of the developmental identity of these structures in fish and terrestrial vertebrates:

> the transcendental anatomists infer, that the vertebrate (if not the invertebrated) animals are constructed according to the same *type* or plan; and that the higher animals, before arriving at their ultimate degree of development, successively run through stages in which their structure is similar to that of animals of less complex organization.[150]

It was, however, quickly pointed out by Cheek that this did not mean that the foetus of a human being actually was a fish, a reptile or a bird at any stage in its development, as not all the organs passed through the same stage of development at the same time. The article went on to refer to Geoffroy's 1829 transformist article from the *Mémoires du Muséum d'Histoire Naturelle*, noting that from 'this gradual process of formation we can understand the production of monsters, some of which have been shown by [Geoffroy] St Hilaire to be caused by stoppage of the development of some of their parts'.[151] Based on the evidence of his articles in the *Edinburgh Journal of Natural and Geographical Science*, Cheek, like his older contemporaries in Edinburgh medical circles, Grant and Knox, was almost certainly a disciple of Geoffroy, subscribing to both his theories of unity of plan and of transformism. It may well have been through either Knox or Grant, the latter a fellow member of the Plinian Society, that Cheek was first exposed to these ideas.

Cheek not only discussed these issues in the pages of the journal he edited, but also in a remarkable dissertation he read before the Royal Medical Society on 29 January 1830. In this dissertation Cheek made a notable contribution to the debate on the origins of human races to be found in the Society's dissertation books. Like most of his contemporaries, Cheek supported the monogenist view, but he went well beyond simply proposing a common origin for the varieties of the human race. He questioned the distinction between species and varieties, and the very existence in nature of species as conventionally understood. He went as far as to propose a common origin for all species. First he tackled the species concept itself, admitting:

> I have not met with an author who can distinguish the species from the permanent variety. And Buffon probably was not so widely inaccurate, when he said 'qu'il n y a pas d'espece dans la Nature', for if an animal be liable to the casual production of varieties, whose new characters are transferable to its offspring, these definitions of species must either be declared to be inexact, or the nonexistence of species must be admitted.[152]

As well as reflecting the views of Buffon, this also accords with Lamarck's opinion on the unreality of species. According to this view, all classification was therefore an artificial creation of natural historians, which were useful only from the perspective of one particular privileged moment in the history of the earth.

Cheek then addressed the question of the distinction between species and varieties. First he noted that the conventional distinction was based on 'the <u>origin</u> of the difference'. If we imagine we can find a natural cause for the difference between two types of living thing – presumably Cheek was thinking principally of the effect of the conditions of life here – we name them varieties, if not, we must assume that the difference has existed *ab initio*, and we name them species. As Cheek himself put it: 'If we fancy we can devise a probable cause for a particular diversity, we name it a variety. If mystery overpowers our subtlety, we name it species'.[153] Having cast doubt on the distinction between species and varieties, Cheek then turned to the question of how both arose. He summarised the various possibilities as he saw them in this remarkable passage:

> All the varieties, as well as all species, may have been transmitted from a single pair, by spontaneous changes; or some physical alterations in the constitution or revolutions of the globe, and its atmosphere may have produced them from a specific type, as some similar changes may have caused the differences of species, whilst previously their several characters were associated under a genus; or lastly the characteristics of permanent varieties, or rather of species may have appeared at once over all those portions of the globe, were the necessary conditions of existence assembled.[154]

Here Cheek gives us three possible theories for the origin of species.[155] First, he considered the possibility that 'spontaneous changes' may be responsible for the diversity of life. This implies that living things have an innate tendency to change their forms over time. Second, he suggested that species may be moulded directly by changes in their conditions of life, so as to generate new species as these change. Third, Cheek considered the possibility that species and varieties came into being simultaneously wherever the conditions were suitable for them. The first model is close to Lamarckian transformism, in which the principle mechanism for the transmutation of species was an innate tendency to increasing perfection, although Lamarck did recognise that changes resulting from changing external conditions could complicate this simple progressive pattern.[156] The second is closer to the theories of Geoffroy, who proposed that directional change in the physical conditions of the earth, in this case in the composition of the atmosphere, led to progressive changes in living things. The third theory is essentially the model proposed by Buffon in his *Époques de la Nature* (1778), where he suggested that as the earth cooled over time, new species arose that were adapted to the cooler conditions, while warmth-loving species became extinct.[157] Although Buffon gave no indication of how he thought this came about, he was clearly proposing not the transmutation of species, but rather the appearance *de novo* of entirely new species. Curiously, Cheek then appeared to dismiss the whole question as a 'mere barren waste of speculation' before giving his own opinion that transmutation came about through the action of a 'Zoological law', which generated variation in living things.[158] Was this an attempt to deflect criticism from his own opinions by denying the validity of any such theorising? It is impossible to say for certain.

Like Grant's, Cheek's model of transformism seems much closer to Geoffroy than Lamarck. Following Geoffroy, Cheek suggested that physical conditions, while not causing adaptive change directly, did exert a selective pressure on variant forms that arose, eliminating those forms which are ill-adapted to the conditions in which they found themselves. In his Royal Medical Society dissertation, he asserted that:

> the climate does not <u>cause</u> the change – but that there is an ultimate Zoological law, that structures have a tendency to change for adaption to new functions – that the indigenous races are adapted to the climates in which they are found – and that, if a race be removed to a new region it will either become adapted to the new functions required, by the powers of organization, or that it will propagate a sickly & imbecile offspring, and ultimately perish.[159]

The transmutations themselves, on which the selective action of physical conditions can operate, occur through the action of a 'Zoological law', which appears to manifest itself as an innate tendency to produce new variations in living things. Once again, this is strikingly reminiscent of Geoffroy's transformist theory. The paragraph quoted from Cheek's essay above is followed by this enigmatic passage:

> The question as to the unity of origin is not necessarily connected with the inquiry as to identity of species. Founded on historical & traditional considerations, the supposition of a single origin for every species does not partake of the nature of a fact to be admitted in Zoology. I believe it to be impossible, and am prepared with my proofs.[160]

Taking 'single' to mean 'separate', this would seem to be a strong affirmation of Cheek's belief that all species have a common origin, and that the contrary opinion was only the result of 'historical & traditional' beliefs obscuring the truth. However, taking 'every species' to mean 'all species', he would seem to be saying the opposite. Taken in the context of what has gone before, the first interpretation appears the more likely, although Cheek seems to have chosen to express himself in a way that is curiously opaque and ambiguous. Of the proofs he had prepared, he sadly had no more to say. Cheek then ended the dissertation on a strangely disconsolate note, remarking, 'I am so much dissatisfied at the mode in which I am obliged to treat this problem which I have selected from a transcendental philosophy that it will be better now to terminate.'[161] Who or what had obliged him to treat the problem in a way that he seemed to find so unsatisfying is unfortunately left to the imagination of the reader. Cheek's dissertation raises as many questions as it answers. What we are not left in doubt about is that Cheek was a convinced transformist.

In the April 1830 issue of Cheek's journal an article was published entitled 'Suggestions on the relation between organized bodies, and the conditions of their existence', which sheds further light on his tranformist opinions. It appeared in a section of the journal called 'Natural-Historical Collections', which was composed of miscellaneous short articles on subjects of natural-historical interest. Like most of the other pieces in this section, the article was anonymous, but was almost certainly by Cheek, given its similarity in tone and language to his Royal Medical Society essay and other articles from the journal that he published under his own name. The 'conditions of existence' mentioned in the title were first defined as 'the external physical agents with which the organized body is in necessary relation, and upon which the integrity and action of its functions depends'.[162] 'Conditions of existence' is a term often associated with Cuvier, for whom it was synonymous with 'final causes', in the sense that the purpose of anatomical structures was to fit the living thing for a specific set of conditions and style of life.[163] The term was also used by Geoffroy, but for him the conditions of existence had become an efficient cause, driving variation and ultimately transmutation.[164] For Cuvier living things were pre-adapted for their destined 'conditions of existence', while for Geoffroy the 'conditions of existence' themselves provided the motor for adaptive change. It is most likely from the latter source that Cheek borrowed the term. The article goes on to explain the relationship between living things and their conditions of existence in the following seven propositions, which merit quotation in full:

1. The development of the process of organization, – a power imposed by the Deity upon matter, – depends upon the conditions of existence.

2. The perfection of organized bodies, or the number and complexity of organs, has a direct ratio with the number of the conditions of existence.

3. All organized bodies possess the power of varying the development of the organs, by addition or subtraction of parts, as changes in the conditions of existence occur.

It is easy to conceive that an organized body can assimilate elements in the form of a new organ, as new functions are required, when we recollect that it is constantly exercising a power of converting inorganic matter into the living emblem of its original form.

4. The characters of organized bodies will be permanent during the continuation of the same conditions of existence which led to their development, and no longer.

5. The more numerous the conditions of existence, the less liable the characters of the organized body to change, and *vice versa*.

6. It has been observed that the older formations of the earth's crust, generally speaking, the less perfect the organic remains they contain. This progressive increase of perfection of organization, would lead us to expect, from the foregoing principle, that, with the advancing age of the earth, the conditions have increased in number; and this seems to be a fact.

7. Adaptation of the law by which organized bodies change with the variation of the conditions of existence; and separation of the functions of relation, and concentration of the vital functions, seems to be the mode of perfection.[165]

Although some of this seems rather obscure, it would appear undeniably to be an unambiguous theoretical statement of the principle that the development of the organisation of living beings and the increasing perfection of living forms over geological time are directly connected to change in their conditions of existence, in a manner broadly in harmony with the theories of Geoffroy. It is worth noting that the idea expressed in the fourth point, which states that living things will remain the same as long as their external conditions remain unchanged, is identical to a concept developed in Geoffroy's transformist article on gavials, published in 1825, and may be derived directly or indirectly from that source.[166] It is furthermore suggested that new organs appear, or are lost, in response to changes in the conditions of life. This is a very similar concept to the principle of the development or disappearance of organs through use or disuse that is central to Lamarck's transformist thought. Lamarck had written in his *Philosophie zoologique* that 'the failure to use an organ, become constant because of the habits which have been adopted, gradually reduces the organs, and finishes by making it disappear'.[167] Conversely, he wrote that 'the frequent use of an organ which has become constant through habit, augments the faculties of that organ'.[168] These changes are then transmitted to offspring. Here as elsewhere, we see Cheek drawing on the work of contemporary transformist thinkers to produce his own synthesis.

The subject of the inheritance of acquired characteristics, so central to Lamarck's theory, was raised by Cheek in a communication read to the Plinian Society on 5 April 1831. This was his last communication to the Society, although his last recorded attendance was in fact not until 3 April 1832. In it he recounted a case where a 'pointer bitch had had her tail mutilated in a trap or by some such accident, in a litter of three pups she produced shortly after they all had tails with a less number of caudal vertebrae than the species in general have'.[169] This is a clear example of the inheritance of acquired characteristics, implicated by both Lamarck and Erasmus Darwin as an important mechanism in the formation of new species. The same subject was raised again in two articles published in March and May

1831 in the *Edinburgh Journal of Natural and Geographical Science*, one probably by Cheek, the other by B. S. Shuttleworth, in which the transmission of mutilations suffered by animals to the next generation was discussed in the context of the theories of Meckel and Johann Friedrich Blumenbach (1752–1840).[170] Cheek was, however, sceptical that a crude modification of an individual animal caused by a single accidental event, such as the loss of a tail, could be transmitted from generation to generation:

> We have, in a former number, stated it to be our opinion, that the characters of an organism are permanent during the operation of the circumstances, internal or external, which produce them, *and no longer*. Whence, the original deficiency of caudal vertebrae being the result of mutilation, a new operation would be required in every successive generation, to continue the character (if it may be so called) in the race.[171]

It can be inferred from the above that Cheek saw the inheritance of acquired characteristics as unnecessary to explain the transmutation of species, and that the increasing complexity of living things over time was both driven and maintained directly by the increasing complexity of the conditions of existence. Here again he deviated from the Lamarckian model of transformism, and was closer to Geoffroy.

Although some transformists in Scotland looked back to eighteenth-century thinkers such as Erasmus Darwin and Buffon, Lamarck and Geoffroy were by far the most important influences in the Edinburgh of the 1820s and early 1830s. Of the two, Lamarck's ideas seem to have been more widely known, although they were often misunderstood. Both his *Philosophie zoologique* and *Histoire naturelle des animaux sans vertèbres* were reviewed by leading Scottish natural historians soon after publication. The latter of these books, which had a first volume almost entirely given over to an exposition of Lamarck's transformist theories, rapidly became an essential reference work for anyone with an interest in invertebrate zoology. The reaction to these books during the 1820s was in general cautiously positive. Even those reviewers who found Lamarck's theories to be excessively speculative generally reserved judgement in the absence of conclusive evidence one way or the other. Evangelical critics, who might be expected to be most hostile towards Lamarck's ideas, were surprisingly circumspect, although attitudes were to harden in this quarter towards the end of the 1820s and in the decades that followed. I have found no evidence that anyone in Edinburgh uncritically accepted Lamarck's model of transformism in its entirety. Both the author of 'Observation on the nature and importance of geology' and Robert Grant claimed to be disciples of

Lamarck, at least if Darwin's account of his conversation with Grant is to be believed. But they both maintained that directional change in physical conditions played a crucial role in determining the direction of transmutation. This is quite different from the Lamarckian model, which relies on an innate tendency to increase in complexity possessed by all living things. Lamarck, in geology a uniformitarian, even denied the reality of directional change in the physical conditions on the earth.

In comparison to Lamarck, Geoffroy's theories were less widely known. This may have been at least in part due to where and how they were published. While Lamarck published complete accounts of his theory in two substantial and widely read works, Geoffroy's transformist speculations were to be found scattered through a handful of journal articles. These were written in a dense and difficult style that was likely to put off all but the most enthusiastic readers. However, a rather more lucid summary of his ideas was published in a translation of a French article in the *Edinburgh New Philosophical Journal* in 1829. It seems likely that Jameson was first exposed to Geoffroy's transformist views through this article, and was so impressed that he decided to give a paper to the Wernerian Society on Geoffroy's theory a few months before the article was published. It seems likely, however, that most of those who were influenced by Geoffroy's transformist speculations came to them by way of his transcendental anatomy. Teachers such as Grant, Knox and Fletcher all drew heavily on Geoffroy and many students must have first heard his name in their lectures. Grant and Knox both knew Geoffroy personally and can be considered his disciples.

There are three Scottish transformist thinkers who were influenced by the theories of Lamarck and Geoffroy and who left reasonably detailed accounts of their ideas in published and unpublished sources: Robert Grant, Henry Cheek and the anonymous author of the 'Observations on the nature and importance of geology', who was almost certainly Robert Jameson. All three integrated the new ideas coming from Paris with an older emphasis on the key role played by directional change in physical condition on the earth's surface inherited from Wernerian geology. Even Cheek, who had little respect for Jameson and whose ideas have the least connection with geology, still depended on directional change in physical conditions on the earth to drive the transmutation of plants and animal species. The idea that conditions of existence drove transmutation was also in harmony with the ideas expressed in the writing of Geoffroy, who saw changes in the composition of the atmosphere as crucial in explaining the history of life. This model of transformism is radically different from those of Lamarck or Darwin, in that it relies on a direct impetus from changing physical conditions rather than by mechanisms that are intrinsic to living things themselves. Nonetheless, although Geoffroy's ideas had more practical impact on the development of transformist thought in Edinburgh, these figures also almost certainly drew inspiration from Lamarck's better known works.

## NOTES

1. Lamarck, *Histoire naturelle*, vol.1, p.317.
2. Ibid., p.26.
3. Ibid., p.331.
4. Lamarck, *Histoire naturelle*, vol.1, p.12.
5. Lamarck, *Organisation des corps vivants*, p.101.
6. Lamarck, *Histoire naturelle*, vol.1, p.43.
7. Lamarck, *Organisation des corps vivants*, p.196.
8. Ibid., p.159.
9. Lamarck, *Histoire naturelle*, vol.1, p.45.
10. Lamarck, *Organisation des corps vivants*, pp.107–8.
11. Lamarck, *Organisation des corps vivants*, p.105.
12. Lamarck, *Philosophie zoologique*, vol.1, p.221.
13. Lamarck, *Organisation des corps vivants*, p.50.
14. Corsi, 'Lamarck: From myth to history', p.12.
15. Lamarck, *Philosophie zoologique*, vol.1, p.74.
16. Lamarck, *Philosophie zoologique*, vol.1, p.75.
17. Lamarck, 'Sur les fossiles des environs de Paris', p.303.
18. Lamarck, *Philosophie zoologique*, vol.2, p.75.
19. Irritability was used to refer to the simple ability of living tissue to contract when touched, as opposed to sensibility, which involved a reaction to a sensation and was mediated by the nervous system.
20. Lamarck, *Histoire naturelle*, vol.1, p.455.
21. Ibid., p.460.
22. See, for example, Jordanova, *Lamarck*, pp.35–6; Burkhardt, 'Inspiration of Lamarck's belief in evolution', p.422 and Laurent, 'Cuvier et Lamarck', p.147.
23. Lamarck, 'Prodrome d'une nouvelle classification des Coquilles', p. 63.
24. Lamarck, 'Sur les fossiles des environs de Paris', p.477.
25. Jordanova, *Lamarck*, p.35.
26. Corsi, 'Lamarck: From myth to history', p. 13.
27. Geoffroy, 'Preliminary discourse' from *Anatomical Philosophy: On the Respiratory Organs*, p.34.
28. Ibid., p.34.
29. Geoffroy, 'First memoir', from *Memoirs on the Organization of Insects*, p.58.

30. Geoffroy, 'General considerations on the vertebra', p.71.
31. It is worth noting at this point that 'unity of plan' in the work of Geoffroy did not take imply that the archetype was an ideal Platonic form existing in the mind of God in the way it did in the writings of, for example, Richard Owen.
32. Appel, *The Cuvier–Geoffroy Debate*, p.130.
33. Geoffroy, 'Recherches sur l'organisation des gavials'.
34. Ibid., p.154.
35. Ibid., pp.154–5.
36. Geoffroy and Serres, 'Rapport fait à l'Académie royale des Sciences sur une mémoire de M. Roulin', p.208.
37. Ibid., p.215.
38. Geoffroy, 'Recherches sur l'organisation des gavials', p.152.
39. Rostand, 'Étienne Geoffroy Saint-Hilaire et la tératogenèse expérimentale', p. 42.
40. Geoffroy and Serres, 'Rapport fait à l'Académie royale des Sciences sur une mémoire de M. Roulin', p.227.
41. Geoffroy, 'Quatrième mémoire', p.79.
42. Geoffroy, 'Palaeontographie', p.82.
43. Geoffroy, *Études progressives d'un naturaliste pendant les années 1834 et 1835*, p.118.
44. Geoffroy and Serres, 'Rapport fait à l'Académie Royale des Sciences sur une mémoire de M. Roulin', p.214.
45. Goulven, 'Le cheminement d'Étienne Geoffroy Saint-Hilaire', p.51.
46. Bourdier, 'Lamarck et Geoffroy Saint-Hilaire', pp. 325.
47. Appel, *The Cuvier–Geoffroy Debate*, p.130.
48. Geoffroy, *Recherches sur de grands sauriens*, p.137 and Geoffroy, *Études progressives*, pp.85–6.
49. Centre National de la Recherche Scientifique, 'Liste des auditeurs du cours de Lamarck'. It should be noted that two Scottish students, Andrew Combe and William Mackenzie, both from Edinburgh, are wrongly counted as English on this website.
50. [Muirhead], Review of Lamarck's *Philosophie zoologique*.
51. Ibid., p.481.
52. Ibid., p.482.
53. Ibid., p.483.
54. Ibid., p.485.
55. See, for example, his discussion of transformist theories in Walker, Notes on natural history lectures (1790), f.163 and Walker, Notes on natural history lectures (1797), vol.9, f.140.
56. Jameson, 'Preface', in Cuvier, *Theory of the Earth*, 1st edn, p.viii.
57. Treuttel & Co. receipts, Jameson papers. Jonathan R. Topham has identified Treuttel & Co. as one of the most significant importers of continental books into Britain in the years following the Napoleonic Wars. See Topham, 'Science, print and crossing borders', in Livingstone and Withers, p.313.

58. Johnston, 'A few remarks on the class Mollusca', p.75.
59. Minutes of the Plinian Society, vol.1, f.57.
60. Anon, Fragments of a translation of Jean-Baptiste Lamarck, *Histoire naturelle*.
61. Rae, *Knox: The Anatomist*, p.109.
62. Secord, 'Edinburgh Lamarckians', p.1.
63. Anon, 'Nature and importance of geology', p.296.
64. Ibid., p.296.
65. It is worthy of note that Jameson himself used the term 'evolved from' here to describe the process by which new species are derived from pre-existing ones. This is a very early use of this term in its modern sense.
66. Anon, 'Nature and importance of geology', p.298.
67. Ibid., p.297.
68. Ibid., p.298.
69. Lamarck, *Philosophie zoologique*, vol.1, p.76.
70. Anon, 'Nature and importance of geology', p.298.
71. Cuvier, *Recherches sur les ossements fossiles*, p.80.
72. Anon, 'Nature and importance of geology', p.299.
73. Ibid., p.297.
74. Jacyna, *Philosophic Whigs*.
75. Ibid., p.138.
76. Darwin, *Autobiographies*, p.24.
77. Grant, 'Observations on the generation of the Lobularia digitata', p.110.
78. Ibid., p.110.
79. Sloan, 'Darwin's invertebrate program', in Kohn, *Darwinian Heritage*, pp.83–4; Grant, *Introductory Essay on the Study of the Animal Kingdom*, p.20.
80. Lamarck, *Histoire naturelle*, vol.1, p.84.
81. Quoted in Sloan, 'Darwin's invertebrate program', p.78; from Tiedemann, *Systematic Treatise on Comparative Physiology*.
82. Grant, 'University of London lectures: Lecture VI. On the organs of support of acephala and echinoderma', p.276.
83. Grant, *Essay on the Study of the Animal Kingdom*, p.17.
84. Ibid., p.18.
85. Ibid., p.12.
86. Ibid., p.17.
87. Rehbock, *The Philosophical Naturalists*, p.50.
88. Richards, 'The "moral anatomy" of Robert Knox', p.399.
89. Powell, *Essays on the Spirit of the Inductive Philosophy*, p.395.
90. Knox, 'Introduction to Enquiries into the philosophy of zoology', p.625.
91. Ibid., pp.190–1.
92. Ibid., p.369.
93. Knox, *Great Artists and Great Anatomists*, p.72.
94. Corsi, *The Age of Lamarck*, p.177.

95. Fletcher, *Rudiments of Physiology*, vol.1, p.12.
96. Ibid., p.12.
97. Ibid., pp.64–5.
98. For a more detailed study of this group of scientific Evangelicals, see Baxter, 'Deism and development', pp.98–112.
99. See, for example, Scott, *Harmony of Phrenology with Scripture*.
100. Grierson, 'General observations on geology and geognosy', p.404.
101. Burns, 'John Fleming and the geological deluge', pp.205–25.
102. Corsi, 'The importance of French transformist ideas', pp.222–4.
103. [Fleming], Review of Lamark, *Histoire naturelle*, pp.403–4.
104. Ibid., p.406.
105. Ibid., pp.406–7. The emphasis on divine power here is typical of evangelical critics.
106. Ibid., pp.406.
107. Ibid., p.408.
108. Ibid., pp.409–10.
109. Cuvier, 'Memoir of M. de Lamarck', pp.14–15.
110. [Fleming], Review of Lamark, *Histoire naturelle*, p.418.
111. Ibid., p.14.
112. Ibid., pp.22–3.
113. Ibid., p.39.
114. Ibid., p.23.
115. [Fleming], Review of Bicheno, 'On systems and methods in natural history', p.320.
116. Ibid., pp.320–1.
117. Anon, 'Of the continuity of the animal kingdom', pp.209–29.
118. Desmond, 'Robert E. Grant', pp.201–2.
119. Pietro Corsi, personal communication. For the original article, see Anon, 'Science: Académie des Sciences', pp.207–8.
120. Wernerian Natural History Society, Minutes of the Wernerian Society, vol.1, f.297.
121. Grant, 'On the structure and nature of the *Spongilla friabilis*', p.283.
122. Geoffroy, 'Organisation des gavials', pp.151–2. The explanation given above is a simplified paraphrase of the rather convoluted French original, which reads: 'Que les décompositions animals, le reformations et les nouvelles compositions se passent dans un même milieu et sous l'action des même agens physique et chimiques, les mutations se reproduisent de la même manière; d'où, à chaque métamorphose, c'est à dire, dans chaque âge, les êtres places sous les influences restent des répétitions exacts les uns des autres. Mais que, tout au contraire, il en soit autrement: des nouvelles ordonnées, si elles interviennent sans rompre l'action vitale, font varier nécessairement les êtres qui en ressentent les effets; chaque fois, c'en est une conséquence toute naturelle, dans le dégrée de leur puissance modificatrice.'
123. Desmond, *The Politics of Evolution*, p.56.

124. Geoffroy. 'Quatrième mémoire', p.79.
125. Grant, *Essay on the Study of the Animal Kingdom*, p.3.
126. Grant, Notes on 'Lectures on comparative anatomy' (1833–34), f.3.
127. Ibid., f.2.
128. Ibid., f.156.
129. Grant, *Comparative Anatomy and Zoology* (syllabus), p.6.
130. Knox, 'Lectures of M. De Blainville', p.137.
131. Rehbock, *Philosophical Naturalists*, p.50.
132. Geoffroy, 'Recherches sur l'organisation des gavials', pp.97–155.
133. Appel, *The Cuvier–Geoffroy Debate*, p.130.
134. For example Knox, *Races of Men*, pp.176, 478.
135. Knox, *Great Artists and Great Anatomists*, pp.211–12.
136. Ibid., p.109.
137. Knox, 'The philosophy of zoology, with special reference to the natural history of man', p.218.
138. Fletcher, *Rudiments of Physiology*, pp.9–10.
139. Ibid., p.12.
140. Ibid., p.12.
141. Ibid., p.12.
142. Ibid., pp.14–15.
143. Minutes of the Plinian Society, vol.2, f.85 recto.
144. Ibid., f.93 verso.
145. For a comprehensive and detailed account of the debate and its impact, see Appel, *The Cuvier–Geoffroy Debate*.
146. [Cheek], 'Review of the recent discussion before the Academy of Sciences in Paris. Part I,– Baron Cuvier's Views', p.37
147. Appel, *The Cuvier–Geoffroy Debate*, p.171.
148. [Cheek], 'On the present state of science abroad', p.116.
149. [Cheek], 'On the natural history of the dugong', p.162.
150. [Cheek], 'On the existence of vascular arches', p.235.
151. Ibid., p.235.
152. Cheek, 'On the varieties of the human race', p.301.
153. Ibid., p.302.
154. Ibid., p.306.
155. Although Cheek doubted the existence of species in nature he continued to use the term in his writings, presumably for want of a suitable alternative. I have followed his example when discussing his theories.
156. Lamarck, *Philosophie zoologique*, vol.1, pp.218–19.
157. Buffon, *Histoire Naturelle, Epoques de la Nature, supplément*, vol.5.
158. Cheek, 'On the varieties of the human race', p.306.
159. Cheek, 'On the varieties of the human race', pp.306–7.
160. Ibid., p.307.
161. Ibid., p.307.
162. [Cheek], 'Suggestions on the relation between organized bodies', p.65.

163. See, for example, Cuvier, *Règne Animal*, vol.1, p.6.
164. See, for example, Geoffroy, 'Anencéphales humains', p.267.
165. [Cheek], 'Suggestions on the relation between organized bodies', p.65.
166. Geoffroy, 'Recherches sur l'organisation des gavials', pp.97–155.
167. Lamarck, *Philosophie zoologique*, vol.1, p.240.
168. Ibid., p.248.
169. Minutes of the Plinian Society, vol.2, ff.90 verso–91 recto.
170. Anon, 'Query on the hereditary transmission of accidental characters', p.173; and Shuttleworth, 'Hereditary transmission of accidental characters', p.301.
171. [Cheek], Editorial comment on 'Query on the hereditary transmission of accidental characters', p.173.

# 6

# The Legacy of the 'Edinburgh Lamarckians'

This chapter will look beyond the early 1830s to evaluate the legacy of transformist debates in Edinburgh in subsequent decades. After the mid-1830s, transformism largely seems to have dropped out of sight and it becomes difficult to identify any significant pro-transformist views expressed in Edinburgh natural history circles. Some of the reasons why this might have been the case are explored in the next section. Despite the apparent eclipse of transformism in Edinburgh, a number of figures who had been teachers or students in the city in the 1820s came to be identified as transformists in later decades, while Robert Grant continued to promote his transformist views of the 1820s in his subsequent career in London, where he held the chair of comparative anatomy at University College London from 1827 until 1874. Robert Knox and Hewett Cottrell Watson, on the other hand, fall into the category of those for whom there is evidence of a belief in transformism only from the period after 1830, while such evidence is lacking from earlier phases of their careers. Knox and Watson, a former student of Jameson, both published works containing assertions of the mutability of species in the decades before the publication of the *Origin of Species*, although neither produced a major work devoted to the subject.

When evolutionary speculation did re-emerge into the light of day as the subject of a book written by an Edinburgh man in 1844 it came from an unexpected source. Rather than a figure from the circle around Edinburgh's professor of natural history, the author was a self-taught natural historian with no formal education in the subject at all. His book became a major bestseller, going through fourteen editions between October 1844 and July 1890, sparking a fierce debate on transformism and bringing the subject to a wide middle-class audience for the first time.[1] This chapter will address the question of whether there was indeed any connection between the transformists of the 1820s and Robert Chambers, the anonymous author of *Vestiges of the Natural History of Creation*.

At the same time as Chambers was researching and writing his great hymn to evolutionary progress, the theory of evolution that was finally to gain widespread acceptance for the transmutation of species was also gradually taking shape in the notebooks of Charles Darwin. Although these theories were to be significantly reworked in later decades as a result of Darwin's own researches and his encounter with the works of Thomas Malthus, these early sketches of his theory show interesting parallels with earlier transformism speculations. Can we see any continuity between the transformism of the late 1820s in Edinburgh and the early musings of Darwin a decade later on his return from the voyage of the *Beagle*? This will be the question addressed by the final section of this chapter.

## THE ECLIPSE OF TRANSFORMISM IN EDINBURGH

After 1832, open advocacy of progressive, gradualist visions of the history of life becomes increasingly rare in Edinburgh natural history circles. Wernerian geology itself found few defenders after the mid-1820s and Robert Jameson, the high priest of neptunism, became more and more an isolated figure among geologists. Cuvierian catastrophism, championed in England by such figures as William Buckland, William Conybeare and Adam Sedgwick, for a time carried all before it. Cuvier had proposed that the geological and fossil records revealed evidence for a sequence of 'worlds', separated from each other by a series of global catastrophes. Both the continuity of the histories of the earth and of life were therefore broken by profound discontinuities. Buckland, for example, suggested in his 1836 Bridgewater Treatise that the history of life on earth had been punctuated by 'revolutions and catastrophes, long antecedent to the creation of the human race', for which there was bountiful evidence in the geological record.[2] In his *Discourse on the Studies of the University* Sedgwick agreed, asserting that 'our globe has been subject to vast physical revolutions'.[3] Sedgwick went on to make clear that the creatures of the new creations that followed these revolutions showed a radical discontinuity with previous forms, and 'though formed on the same plan, and bearing the same marks of wise contrivance, oftentimes [are] as unlike those creatures which preceded them, as if they had been matured in a different portion of the universe and cast upon the earth by the collision of another planet'.[4]

Catastrophism, implying as it did a more or less complete turnover of flora and fauna at the time of each catastrophe, was fundamentally incompatible with the picture of the gradual development of life driven by changes in the physical conditions on the surface of the earth that Wernerian geology had suggested to many earlier geologists. Ironically,

Jameson had done much to promote catastrophist ideas as the editor of successive editions of Cuvier's *Theory of the Earth*, for which he also provided extensive notes. The picture of the history of the earth presented in the *Theory of the Earth* may not have seemed to Jameson to challenge the Wernerian picture of gradual, progressive change in living things, as Cuvier himself admitted that marine organisms had undergone transmutations brought about by changes in the properties of the medium in which they lived. There is a striking statement of this in Jameson's translation for the fifth edition of *Theory of the Earth* (1827), where, closely following the original French text, it is noted: 'There has, therefore, been a succession of variations in the economy of organic nature, which has been occasioned by those of the fluid in which the animals lived, or which at least corresponded with them; and these variations have gradually conducted the classes of aquatic animals to their present state.'[5] Despite this, the majority of British geologists interpreted the obvious, radical discontinuities in the fossil record as evidence that an entire world of living things had been swept away and replaced with a new creation. Hugh Miller, one of the leading Scottish advocates of discontinuity in the history of life, was to write: 'The curtain drops at his command over one scene of existence full of wisdom and beauty – it rises again, and all is glorious, wise and beautiful as before, and all is new.'[6]

As the quotation above makes clear, it was not just Miller's catastrophist views on the history of life that led him to reject any slow transformation of life over geological time, but also the evangelical faith that underlay them. Miller was utterly opposed to the idea of gradual, progressive development, which he saw as making redundant God's power to create new species by supernatural intervention, leaving him without a direct guiding role in his own creation. Miller's friend and fellow evangelical David Brewster wrote in a review of *Vestiges of the Natural History of Creation* that '[i]f it be a received truth that the Creator has repeatedly interposed in the government of the universe, and displayed his immediate agency in miraculous interpositions, it is an insult to any reader to tell him that that being slumbers on his throne'.[7] The conclusion to be drawn was clear. As Miller asserted in his *Old Red Sandstone*:

> There is no progression. If fish rose into reptiles, it must have been by sudden transformation; – it must have been as if a man who had stood still for half a life-time should bestir himself all at once, and take seven leagues at a stride. There is no getting rid of miracle in this case[.][8]

The Evangelical Party within the Church of Scotland numbered many prominent natural historians and natural philosophers among its ranks, including Miller, Brewster and John Fleming. Some of these Evangelical figures, such as Fleming and Brewster, had been close associates of Jameson in the 1820s. Increasing Evangelical militancy in the decades before their definitive split with the Established Church to form the Free Church of Scotland in the Disruption of 1843 had a profound influence on cultural developments in the period, not least in natural history. In the two decades leading up to the publication of *Vestiges of the Natural History of Creation* in 1844, attacks against transformism from Edinburgh natural historians came almost exclusively from among the ranks of the Evangelicals, and even after the publication of that book led to more widespread condemnation of transformist ideas they very much led the charge in Scotland. A very early Evangelical response to Lamarckian transformism comes from the pages of the *Memoirs of the Wernerian Society*. This took the form of a paper by the Evangelical minister James Grierson, given to the Society in February 1824, which dismissed transformism as mere fanciful speculation. This was to be a principal mode of attack for the Evangelical opponents of transmutation.

Another Edinburgh-based natural historian, member of the Wernerian Society and minister of evangelical opinions who expressed his views on Lamarck in print was James Duncan (1804–61), who wrote a 'Memoir of Lamarck' to accompany a volume on 'Foreign Butterflies' that he wrote for William Jardine's *Naturalist's Library* (1837). His approach to Lamarck's transformism bears comparison to that of Fleming during the previous decade, although Duncan did not seem to have been prepared to accept that Lamarck's theories had any concrete evidence in their favour. He admitted to finding it 'difficult, indeed, to conceive how Lamarck could advance a theory so utterly opposed to observation and probability, and at the same time succeed so effectually in convincing himself of its truth'.[9] Despite the fact that Lamarck's speculative theories were contradicted by the evidence, Duncan nonetheless admitted that 'they merit attention as the production of a mind remarkable for originality and penetration, as well as for extensive and varied knowledge'.[10] Duncan seems to have been thoroughly familiar with the first volume of Lamarck's *Histoire naturelle des animaux sans vertèbres*, which he paraphrased at length, while his scientific criticisms of Lamarck's transformism seem largely to have been derived from volume 2 of Lyell's *Principles of Geology* (1832). Like so many others, he also repeated Cuvier's misrepresentation of Lamarck's theory as one in which 'efforts and desires engender organs'.[11] It is evident that Duncan also derived most of his biographical information on his subject from Cuvier's mean-spirited eulogy to Lamarck.[12] For Duncan, however, Lamarck's transformism was not just bad science, but was also theologically unacceptable, being 'at once absurd and impious – alike opposed to reason and religion'.[13] His critique of Lamarck on theological grounds is based on Duncan's typically evangelical argument that his theory denies God a continuing role in the world after the creation:

> While thus admitting the existence of the Deity, any direct interference in the affairs of the universe is wholly denied to him. His sovereignty is reduced to a mere nominal supremacy, as he is supposed to take no care or thought for the worlds which he authorized or permitted to be created, and can have no sympathy for the creatures which inhabit them.[14]

This critique was to be typical of evangelical responses to transformism in the 1840s, where we find very similar sentiments expressed by Evangelical natural philosophers and natural historians such as David Brewster and Hugh Miller in their critiques of Robert Chambers' *Vestiges of the Natural History of Creation* (1844).[15]

Although John Fleming seems to have supported a progressive history of life on earth in his *Philosophy of Zoology* of 1822, by 1829 he could completely deny that there was any evidence for the progressive appearance of the different orders of animals in a review of J. E. Bicheno, *Systems and Methods in Natural History* (1827), published in the *Quarterly Review*. Fleming's estrangement from Jameson and rejection of Wernerian geology and the progressive view of the history of life raises many questions, although it certainly bore some relation to his religious beliefs. Fleming was a deeply religious man and a minister of the Evangelical Party of the Church of Scotland, and, after the Disruption of 1843, of the Free Church of Scotland. Like many Evangelicals with scientific interests he was horrified by the use made of geological and biological theories in the *Vestiges of the Natural History of Creation*, which he saw as an appalling assault on the principles of true religion. Based on the testimony of his inaugural address as professor of natural history at the Free Church's New College in 1850, he had clearly come to believe that the dangerous 'development hypothesis' outlined in that book had its roots in the Wernerian geology favoured by Jameson and his associates, including Fleming himself in his younger days.[16] Fleming's personal relationship with Jameson also seems to have deteriorated over the years. Jameson was a complex and difficult character, who succeeded in alienating many of those he had dealings with during his long career through his high-handed manner and arbitrary behaviour. Fleming's later estrangement from Jameson is quite evident in a quotation found in John Duns' memoir of Fleming, only published in 1859 after the deaths of both men, where Fleming is quoted as describing Jameson as 'irregular, cold, and distant'.[17] Whatever the reasons, Fleming in later years became one of the most implacable enemies of transformism and progressivism in British geology.

The rise of catastrophist geology and the increasing militancy and influence of the Evangelical Party of the Church of Scotland in the long run up to the Disruption of 1843, which tore the Church of Scotland apart and gave birth to the Free Church of Scotland, certainly provided an increasingly hostile climate for transformationist ideas in the 1830s and 1840s. The personal fortunes of individuals within the Edinburgh natural history community and the trajectories of their careers may also have a strong bearing on the matter. By the early 1830s, some of the key figures I have identified as promoters of transformist theories were no longer in Edinburgh. Robert Grant, the transmutationist who has left more evidence than any other Edinburgh figure for his beliefs in the 1820s, was in London. The fortunes of others had declined significantly and at least one important figure, Henry Cheek, was dead by 1833. Jameson clearly took a strong interest in the transmutation of species in the late 1820s, inspired by Grant and his own encounter with the theories of Geoffroy Saint-Hilaire. However, it never seems to have been a central preoccupation for him, and he may have simply settled back into the study of mineralogy, always his main preoccupation, without the stimulation provided by younger colleagues such as Grant. In any case, the transmutation of species did not occupy a pivotal place in contemporary natural history, but was one question among many for Jameson and his students. The question may have simply faded into the background when no successors to Lamarck and Geoffroy emerged to reanimate debate on the subject in the 1830s. By the early 1830s, the rise of catastrophist geology and the concerted opposition of influential Evangelical natural historians would seem to have made gradualist, developmental theories of the history of life on earth appear increasingly untenable and support for such theories, at least in public, died away. The subject of evolution was, however, to resurface dramatically with the publication of *Vestiges of the Natural History of Creation* in October 1844.

## VESTIGES OF THE NATURAL HISTORY OF CREATION

The publication of *Vestiges* radically changed the terms of the discourse on transformism in Britain and stimulated a lively public debate on the subject. The book appealed to a wide readership, including a large number of middle-class readers who may not previously have taken an interest in such arcane matters. It presented a distinctive theory of transformism set within an all-encompassing vision of universal progress. Its anonymous author was Robert Chambers (1802–71), a well-known Edinburgh journalist and publisher. His magnum opus was an extraordinary publishing success. The first two editions of 750 and 1,000 copies respectively sold faster than the author or publisher had dared to hope, and the third edition of 1,500 copies sold out on the day of publication.[18] In this section I will be asking whether any connection can be discerned between the Edinburgh transformists of the 1820s and early 1830s whose ideas I have been exploring in earlier chapters and the appearance of Chambers' sensational transformist work in 1844. Was it entirely coincidental that the author of *Vestiges* hailed from Edinburgh, or is there some link to be discerned between the transformist speculations of Edinburgh natural historians and medical men in earlier decades and the spectacular emergence of transformist ideas into the public sphere with the publication of *Vestiges*?

Robert Chambers and his elder brother William were born in Peebles in the early years of the nineteenth century, the sons of a small-scale cotton manufacturer. Both Robert and William went into bookselling, at first on a very modest scale, but they soon branched out into publishing their own books. William was an astute businessman, and W. & R. Chambers, founded in 1832, soon became a major Scottish publishing company, producing such innovative publications as *Chambers's Encyclopaedia* and *Chambers's Edinburgh Journal*. In the early days, Robert wrote a large proportion of their publications himself. He and his brother made an excellent team, William taking care of the running of the business, while Robert dealt more with the literary and journalistic side of things.[19] Robert had originally been principally interested in Scottish history and traditions. Among his earliest works were *Illustrations of the Author of Waverley* (1822), *Traditions of Edinburgh* (1824) and *Rhymes of Scotland* (1826). However, as his position in society became more secure in the 1830s, he turned his attention increasingly towards natural history. It was this new interest that was ultimately to lead him towards the formulation of his 'development hypothesis', which found expression in *Vestiges*.

Although the whole of *Vestiges* preaches the doctrine of universal progress by means of the development hypothesis, only Chapters 12 to 14 of the nineteen chapters that constituted the first edition actually addressed transformism directly. The first two chapters deal with the formation of the earth and the other bodies of the solar system in accordance with the nebular hypothesis recently popularised by the University of Glasgow's professor of practical astronomy, John Pringle Nichol (1804–59), in his *Views of the Architecture of the Heavens in a Series of Letters to a Lady* (1837).[20] The next nine present a potted history of the earth and of life that is progressive but not specifically transformist; it would in itself have been familiar and entirely acceptable to such grandees of the geological establishment as William Buckland and Adam Sedgwick. After this come the three chapters in which Chambers gave an exposition of his transformist theories. He then devoted a chapter each to William Sharp MacLeay's eccentric quinary system of taxonomy, a progressivist interpretation of human history, the mental powers of animals, some vaguely natural-theological musings on the purposes of the animal creation, and finally a conclusory note. I will therefore be concentrating my attention on the content of the three chapters that deal directly with transformism. When I refer to *Vestiges*, it is to be understood that I mean the first edition of 1844. Chambers made significant changes to later editions in order to meet the objections of his critics, but these are not directly relevant to my argument here.

Despite the marked evolution of the book from one edition to the next, the essential core of its argument remained the same. Chambers believed that life had first come into being through spontaneous generation, and continued to do so to the present day. His model of the history of life, like Lamarck's, was one that depicted numerous parallel lineages all at different phases of development. Chambers saw in this the working out of a set of natural laws established by the deity. For him the spontaneous generation of life was analogous to the law-governed formation of the globe from a cloud of 'nebulous matter', as proposed in the nebular hypothesis popularised by Nichol; 'one set of laws produced all orbs and their motions and geognostic arrangements, so one set of laws overspread them all with life'.[21] To explain the development of increasingly advanced forms of life, Chambers developed an analogy based on the normal process of gestation. He suggested that every species followed a common developmental path during foetal development up to a given point, at which the species diverged and followed its own track to the adult form. To support his argument, he drew on comparative anatomy, pressing into service the well-worn example of the vertebrate limb, to demonstrate the common basic design that underlay the body plans of all animal species. All that was required for a new, more advanced, species to arise was for that common path of foetal development to be extended before branching off occurred: 'To protract the *straightforward part of the gestation over a small space* – and from species to species the space would be small indeed – is all that is necessary.'[22] Whether the development of the foetus is extended further than normal or was cut short depended on the conditions to which the mother was exposed. In the following passage, Chambers used the example of congenital malformations of the human heart to illustrate the effect of impoverished conditions on living things before showing how the same process in reverse could lead to progress from a lower to a higher form:

> A human foetus is often left with one of the most important parts of its frame imperfectly developed: the heart, for instance, goes no farther than the three chambered form, so that it is the heart of a reptile. There are even instances of this organ being left in the two-chambered or fish form. Such defects are the result of nothing more than a failure of the power of development in the system of the mother, occasioned by weak health or misery. Here we have apparently a realization of the converse of those conditions which carry on species to species, so far, at least, as one organ is concerned. Seeing a complete specific regression in this one point, how easy it is to imagine an access of favourable conditions sufficient to reverse the phenomenon, and make a fish mother develop a reptile heart, or a reptile mother develop a mammal one.[23]

So what did Chambers know about earlier transformist thinkers and what might his sources of information on them and their theories have been? Since he included a critique of Lamarckian transformism in his book, we will pay particular attention to the possible sources of his information on Lamarck's theories. As Chambers made significant use of works on embryology and comparative anatomy in developing his theory we will also explore possible connections between the transcendental anatomy that was so influential in Edinburgh medical circles in the 1820s and 1830s and Chambers' theories; as we have seen, espousal of the transcendental anatomy of Geoffroy in Edinburgh was often also accompanied by a sympathetic attitude towards his transformist speculations. Finally, we will explore the influence of the leading Edinburgh phrenologist and friend of Chambers, George Combe, and his popular and influential book, the *Constitution of Man* (1828 and numerous subsequent editions), which is everywhere apparent in *Vestiges*. We should then be able to reach some tentative conclusions regarding whether the re-emergence of transformism in the mid-1840s in Edinburgh, the home of a notable group of transformists in earlier decades, was coincidental, or whether there was indeed some direct or indirect connection.

Before seeking answers to the question of where Chambers found the building blocks for his theory, we need to establish how much he knew about the theories of earlier transformists and whether these ideas were transmitted to him directly or through writings of secondary sources in Edinburgh or beyond. From the evidence available it seems unlikely that he had himself read the works of Geoffroy or Lamarck, or any of the other continental sources of the ideas he used. As Secord has established, Chambers' 'awareness of continental authors, such as E. Geoffroy Saint-Hilaire, Lamarck, E. Serres, Friedrich Tiedemann and Karl Ernst von Baer, seems to have been entirely at second hand'.[24] In order to establish what Chambers did in fact know about earlier transformist theories and from which sources he gained this knowledge, we will first examine the evidence of *Vestiges*, before turning to Chambers' earlier writings.

Chambers' theory as expressed in *Vestiges* was radically different from that of either Lamarck or Geoffroy. Unlike Geoffroy's transformism, it assumed that transmutation occurred along a fixed developmental track, rather than depending on the influence of changing conditions. Although both theories drew on embryology, Geoffroy saw the production of 'monsters' resulting from deviations from the normal developmental pathway as crucial, while Chambers considered that the production of new forms came about through a prolongation of normal foetal development. His theory had somewhat more in common with Lamarckism, as both assumed an inbuilt tendency towards increasing perfection in living things. However, Lamarck's theory did not depend on embryology and relied rather on changes to the body of the adult form being transmitted from generation to generation. Chambers also rejected absolutely the second strand of Lamarck's theory, which suggested that new habits derived from new needs could modify the bodies of living things, and that these changes could be transmitted to offspring. Indeed, from the evidence of *Vestiges* it would be easy to believe that this was the only part of Lamarck's theory that he was aware of, and even then he seems to have viewed it through the distorting lens of the writings of Lamarck's earlier critics.

That Chambers knew something about Lamarck's theory is undeniable. He devoted a sizable passage in his chapter on his 'hypothesis of the development of the vegetable and animal kingdoms' to a critique of Lamarckian transformism. However, the reader who expected a sympathetic account of Lamarck's theory would be sorely disappointed. When we read Chambers' exposition of Lamarck's theory, we find the same contorted caricature as can be found in the writings of Julien-Joseph Virey, Georges Cuvier, John Fleming and James Duncan. Chambers presented Lamarck's theory as implying that animals willed themselves to evolve in order to meet immediate physical needs. According to Chambers, in Lamarck's theory 'one being advanced in the course of generations to another, in consequence merely of its experience of wants calling for the exercise of its faculties in a particular direction, by which exercise new developments of organs took place, ending in variations sufficient to constitute a new species'.[25] Based on this skewed account, Chambers concluded that Lamarck's theory was 'obviously so inadequate to account for the rise of the organic kingdoms, that we can only place it with pity among the follies of the wise'.[26] In the circumstances, it seems highly unlikely that Chambers had read any of

Lamarck's works, and more than likely that he had gathered his information from one or more of his critics. However, he does not cite any sources for these opinions in *Vestiges*.

One plausible answer to the question of where Chambers got his information on Lamarck from emerges from the pages of *Chambers's Edinburgh Journal* for 26 September 1835, in the form of an article entitled 'Popular information on science: Transmutation of species'. While the article is without attribution and may not be by Chambers himself, it would still have had to pass his editorial scrutiny. Even if Chambers did not write this article, which it is very likely he did, he could hardly have avoided reading it. This article contains a number of references to Charles Lyell's critique of Lamarck, which takes up a large part of volume 2 of Lyell's *Principles of Geology*. Chambers included a lengthy quotation from this work, in which Lyell devoted considerable space to an extremely thorough refutation of transformism.[27] Indeed, the article is largely a summary of the opinions on transformism given in Lyell's book. The picture of Lamarck's theory presented here therefore bears little relation to the one propounded by Chambers in *Vestiges*, because Lyell's critique was radically different, and much more sophisticated. Far from simply repeating Cuvier's assertion that Lamarck had believed that 'it is the desire of flying that has converted the arms of all birds into wings',[28] Lyell gave an essentially fair and accurate account of Lamarck's theory.[29] He seems to have based his analysis largely on a careful reading of the *Philosophie zoologique*. Despite the obvious differences, Chambers may still have had Lyell's account at the back of his mind when he wrote his critique of Lamarck in *Vestiges*; it may not be a coincidence, for example, that both chose the example of the development of webbed feet in aquatic creatures to support their rather different arguments.[30]

Nonetheless, familiar as he may have been with Lyell's critique of Lamarck, this cannot be the source of his belief that Lamarck considered that the 'wants' of animals drove the transmutation of species, as this distortion of his theory is not to be found in the *Principles of Geology*. These ideas may plausibly have come from Fletcher, who we know discussed and rejected Lamarck's theory in his *Rudiments of Physiology*, since Fletcher criticised Lamarck on similar grounds, as we have seen in Chapter 5. Fletcher's critique was, however, rather more nuanced than the one found in *Vestiges*. The translation of Cuvier's 'Memoir' of Lamarck, which appeared in the pages of the *Edinburgh New Philosophical Journal* in 1836, or Fleming's *Philosophy of Zoology* (1822) are also both plausible sources, but it is impossible to determine this with any degree of certainty. This particular caricature of Lamarck's theory seems to have had sufficient currency in natural history circles in the 1830s that there are any number of possible sources.

The 1835 article on the transmutation of species was not the only piece from the *Chambers's Edinburgh Journal* that dealt with the transmutation of species. In November 1832 there appeared an article entitled 'Natural history: Animals with a backbone', in which the transformism of Erasmus Darwin was discussed. The article noted Darwin's belief in spontaneous generation, and that all life had progressed from very simple primordial forms. However, the only one of the several mechanisms proposed by Darwin that is mentioned is the production of new species through hybridisation. Chambers, if he was indeed the author, noted that hybridisation could never give rise to fertile offspring; otherwise 'the world would become a scene of hideous monstrosity'.[31] Darwin's theories were in general dismissed as 'baseless doctrines'. According to the author of the article, 'views like these can never be entertained by healthy minds, and it required little reflection to dispel such absurd theories'.[32] In 1840, Darwin made another appearance in the *Journal*, this time in an article on 'The life and poetry of Darwin'. Here again, Darwin's theories were again rejected as 'too hypothetical, and even fantastical, to stand the test of sober and close examination'.[33] The discussion of spontaneous generation in these articles, may, however, have brought the idea to the attention of Chambers, in whose later theories it would take an important place. Other articles in Chambers' *Journal* dealt with theories that were to become important components of the transformist doctrine advocated in *Vestiges*. In an article on 'Natural history: Monkeys, apes, and orang-outangs', published in December 1832, the reader was introduced to the idea of unity of plan in comparative anatomy.[34] A few years later, in October 1837, an article on 'Popular information on science: Third ages of animal life', informed its readers that 'the system of organic being was progressive, and graduated regularly onwards from the simplest forms up to those of a higher and more complex character'.[35]

The prevalence of articles of this nature in the *Journal* has prompted Secord to suggest that 'a careful reader of the *Journal* would have been familiar with much of *Vestiges* before it appeared'.[36] It does seem more than likely that a significant proportion of Chambers' ideas were ultimately derived from the extremely diverse body of material he had had to digest in his role as author and editor for the *Chambers's Edinburgh Journal*. There is, however, nothing to suggest that his knowledge of earlier transformist thinkers was transmitted to him directly from the writings of any of the Edinburgh transformists we have examined in earlier chapters, or through any personal acquaintance with them or their ideas. There is certainly no evidence from his surviving correspondence or from published sources that he personally knew or corresponded with any of these earlier figures. It does seem, however, that some of the central elements from which Chambers constructed his model of transformism do derive from published sources close to Edinburgh's medical transformists of the 1820s and 1830s, and it is to these that I turn now.

As we have seen above, Chambers' theory depended to a considerable extent on ideas derived from comparative anatomy and embryology. What were the sources of these ideas and concepts that he wove into his theory? Fortunately for the historian, Chambers cited a great many of his sources in the text of his book. It is therefore easy to establish that for the underpinnings of his theory in comparative anatomy and embryology he drew on the work of seven medical men who wrote on physiology and anatomy: William Benjamin Carpenter, Robert Bentley Todd, Martin Barry, Leonard Horner, Allen Thomson, Percival Barton Lord and John Fletcher. Of these seven, all but Todd had studied medicine at the University of Edinburgh in the first four decades of the nineteenth century. While none of these figures advocated transformism, Chambers drew on aspects of their work in developing his own transformist theory. Of these seven, however, Chambers relied overwhelmingly on the work of only three: Percival Barton Lord (1808–40, studied at Edinburgh 1832–34), John Fletcher (graduated 1816) and William Benjamin Carpenter (graduated 1839).

The concept of unity of plan, promulgated in Edinburgh by the disciples of Geoffroy in the late 1820s and 1830s, was much in evidence in *Vestiges*. In his chapter on the 'hypothesis of the development of the vegetable and animal kingdoms', Chambers stated:

> While the external forms of these various animals are so different, it is very remarkable that the whole are, after all, variations of a fundamental plan, which can be traced as a basis throughout the whole, the variations being mere modifications of the plan to suit the particular conditions in which each particular animal has been designed to live.[37]

Chambers does not cite any authority for these ideas, aside from a fleeting reference to Louis-Jean-Marie Daubenton's observations on the constancy in the number of neck vertebrae in mammals. However, he would have readily found these ideas in Fletcher's *Rudiments of Physiology* (1835–7) (which we know from citations elsewhere in *Vestiges* he had read), where an exposition of unity of plan takes up most of section II of part 1 of the book. There Fletcher, who, as we have seen in Chapter 3, was an enthusiastic disciple of Geoffroy, stated confidently that 'however different may seem, both in their anatomical and physiological relations, the organs of the higher and those of the lower tribes, if not of plants, certainly of animals, they are both essentially the same'.[38]

Embryology also played a central role in Chambers' theory of the transmutation of species, as it had done in Geoffroy's. However, there the similarity ends. For Geoffroy it was a disruption of foetal development that led to the production of new species. For Chambers the appearance of a new species was simply an extension of the normal process of gestation. As he put it in *Vestiges*, 'the production of new forms, as shewn in the pages of the geological record, has never been anything more than a new stage of progress in gestation'.[39] For Geoffroy the physiological stress caused to organisms by degrading conditions during foetal development spurred transmutation forward, while for Chambers the opposite was true; hostile conditions could only lead to degradation, while favourable conditions were necessary for progress. In his exposition of the role of embryology in evolution Chambers also drew on recapitulation theory as evidence for his theory. This theory, which is sometimes referred to as the 'Meckel–Serres Law' after two of its early proponents, Johann Friedrich Meckel and Étienne Serres, had been championed and popularised by Geoffroy. It suggested that the foetal development of animals recapitulated their evolution of the species over geological time. Chambers noted that 'each animal passes, in the course of its germinal history, through a series of changes resembling the *permanent forms* of the various orders of animals inferior to it in the scale'.[40]

Chambers quoted at length from Percival Barton Lord's *Popular Physiology* to demonstrate that the brain of the human foetus passes through the forms of the brains of all the lower animals, in order of perfection of organisation, before arriving at the adult human form.[41] Lord did not interpret this in a transformist manner, but rather stated that 'man, considered merely as an animal, is, by his organization, superior to every other being; and that, in the growth of a single individual, Nature exhausts, as it were, the structure of all other animals before she arrive at this her *chef-d'œuvre*'.[42] Later in the same chapter Chambers cited Fletcher's *Rudiments of Physiology* to much the same effect as Lord. He quoted Fletcher as noting that 'as the brain of every tribe of animals appears to pass through the types of all those below it, so the brain of man passes through the types of every tribe in the creation'.[43] Fletcher's book seems to contain the germs of a great many of the ideas that appear in *Vestiges*. His implicit link between unity of plan and the development of the individual organism, such that 'the splendid human organism itself consists merely of the same organs, regarded fundamentally, as exist in the polype, the differences consisting chiefly in their different degrees of elaboration', was almost certainly of great significance in the development of Chambers' ideas.[44] Chambers himself acknowledged that the key concept that an extension of the period of gestation leads to the production of a higher form of animal had been 'suggested to me by, in consequence of seeing the scale of animated nature in Dr Fletcher's Rudiments of Physiology'.[45] In fact the idea is even more clearly expressed in the following passage from Fletcher's book, where he stated 'that all tribes of animated nature, with respect to their several organs, start, as it were together, that the germ of each of their organs is in all the same, and that they subsequently differ from each other only or chiefly in their arriving at their appointed goal sooner or later'.[46]

We know that Fletcher was emphatically not a transformist, but he did discuss transformism at length in *Rudiments of Physiology*, as we have seen in Chapter 3. What is more, he linked transformism explicitly with the theory of recapitulation in a way that is strikingly similar to Chambers' theory, although while Fletcher concluded in a footnote that '[t]he system then which would establish that "men and toads" differ only in their greater or lesser development is not tenable', the idea that progress along a single developmental track for all animals was the key to explaining the progressive history of life became the backbone of Chambers' transformism.[47] With some modifications, derived principally from Carpenter, the exposition of the concept of unity of plan and the theory of recapitulation found in Fletcher's book would seem to be largely sufficient to explain the genesis of the central core of Chambers' theory of transformism.

Although Chambers does not credit Carpenter with the idea, Secord has suggested that it was through his writings that Chambers became aware of the theories of Karl Ernst von Baer (1792–1876), according to whom the foetus did not resemble the adult forms of lower animals, but rather their forms at a particular moment in foetal development. Foetal development therefore became a process of differentiation. Von Baer's theory had, in fact, been introduced to the English-speaking world by Martin Barry (1802–55), who had graduated from Edinburgh in 1833.[48] (Chambers also cites Barry in *Vestiges*, although not in this context.) We also know from the correspondence between Chambers and Alexander Ireland, who acted as his intermediary with the publisher to maintain his anonymity, that Carpenter acted as a paid consultant on later editions of *Vestiges*, although not the first one.[49] The version of the theory derived from Carpenter is represented in the diagram from *Vestiges* shown in Figure 6.1, which, as Dov Ospovat has pointed out, is a slightly modified version of a diagram that appears in Carpenter's *Principles of General*

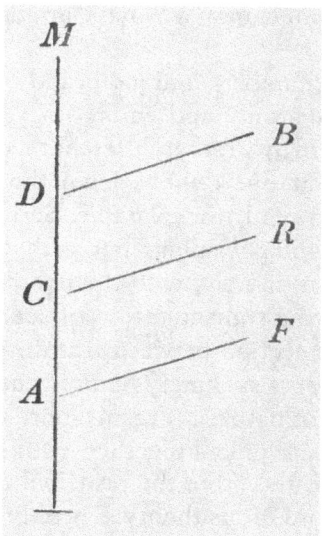

**Figure 6.1** Diagram showing the process of differentiation during development, from Robert Chambers' *Vestiges of the Natural History of Creation*, p.212. Yale University, Harvey Cushing/John Hay Whitney Medical Library.

*and Comparative Physiology.*[50] In this figure F, R, B and M represent the adult forms of fish, reptiles, birds and mammal respectively. All follow a common pathway of foetal development as far as A, where F diverges while R, B and M continue on the same track until C, where R diverges, and so on. For Chambers, all that was necessary for a new species to arise was for the foetus to remain on the common developmental path for longer than normal before diverging.

It is possible that Chambers' debt to Carpenter goes deeper than this, however. In a chapter 'On the evidences of design presented by the structure of organised beings' in his *Principles of General and Comparative Physiology*, we find Carpenter speculating that 'the same Almighty *fiat* which created matter out of nothing, impressed upon it one simple law', which would not only establish the cosmic order of the stars and planets, but would 'people all these worlds with living beings of endless diversity of nature, providing for their support, their happiness, their mutual reliance, ordaining their constant decay and succession, not merely as individuals but as races, and adapting them in every minute particular to their dwelling'.[51] The resemblance of Carpenter's 'one simple law' to Chambers' universal law of progress is striking, and the doctrine of the primacy of natural law that, as we will see below, he inherited from his friend, the phrenologist George Combe, may well have been reinforced by his reading of Carpenter.

We have seen how Chambers' unique brand of transformism drew on ideas on comparative anatomy and embryology that had also helped to inspire and support Geoffroy's rather different theory. However, Chambers only knew these ideas at one remove from the original sources. There is no suggestion he had read, for example, Serres, Meckel or Geoffroy. Many of the authors of the physiology books that Chambers had read did not discuss transformism directly, with the notable exception of Fletcher. While Fletcher rejected the transmutation of species, in his exposition of it he made a crucial connection between transformism, unity of plan and recapitulation theory that is strikingly reminiscent of Chambers' theory of the 'universal gestation of nature'. It seems more than likely that the ideas that Fletcher presented and linked together, with some modification based on his reading of Carpenter, were the principal materials that Chambers used to formulate the core of his theory. These influences provide a tangible if indirect link with the Geoffroyan transformism of Grant and Cheek. Like Chambers' rather different theory, their models of the transmutation of species also emerged in close association with the new transcendental anatomy championed by Geoffroy, which exerted a powerful influence in Edinburgh medical circles in the 1820s and 1830s.

There was one further important influence on Chambers, without which it is impossible to understand the roots of his theory or his motivation for developing it. This influence was also close at hand in Edinburgh, and was transmitted to Chambers through the work of George Combe, author of the *Constitution of Man*. Chambers appears to have become acquainted with Combe and his philosophy in the mid-1830s. In a letter to Combe in December 1833, he described himself as 'not myself altogether ignorant of phrenology, or altogether a sceptic'.[52] He first wrote about the subject in an article for the *Chambers's Edinburgh Journal* of 4 January 1834 entitled 'Is ignorance bliss?', in which he warmly recommended the *Constitution of Man* to his readers.[53] Before publishing the article he sent a proof to Combe, who wrote back to say:

> Considering that you are unacquainted with Phrenology, you have caught a good deal of the spirit of the book, but have not penetrated fully the principle of it, and no one can do this, or be competent to judge of its real importance, who has not the conviction from observation of Phrenology being true.[54]

By late 1835, Chambers had set aside his uncertainty about phrenology. In a letter to Combe dated 25 December of that year, he admitted that he had:

in particular been impressed with the truth of the metaphysical department of the science, and with your singularly excellent work, the Essay on the Constitution of Man, that, in writing upon human nature, I cannot now do otherwise that [sic] employ this philosophy both as a system of mind and of morals.[55]

In the same letter, however, he expressed caution regarding openly broadcasting his adherence to phrenology due to the controversy surrounding the subject. Instead, he thought it better to introduce phrenological principles to his readers by stealth rather than openly declaring their source. Combe did not agree, and wrote to him twice in November 1835 on the subject, on the second occasion saying:

> I am not disposed to dispute that the philosophy of mind developed by Phrenology is, as you say, 'the point of the wedge which is yet to rend asunder the mass of philosophical heathenism'; & I am willing to see it driven by all hands who are disposed to give it a blow. But agreeing with you in this estimate of it, I am anxious that it should be known, when used, to be the Phrenological wedge.[56]

Chambers, however, was not to be persuaded, and continued to refer to phrenology cautiously and infrequently in his published work, a policy that he also extended to the anonymously published *Vestiges*, although, as we will see, it was steeped in phrenological ideas, and Combe's doctrine of universal progress was at the very heart of the work.

Combe was the most influential phrenologist in Britain in the middle decades of the nineteenth century. He had been a successful Edinburgh lawyer until he retired to devote himself entirely to phrenology in 1833. An early convert to this would-be science, his first contact with the subject was through an extremely negative critique of phrenology written by John Gordon that he had read in the June 1815 number of the *Edinburgh Review*. At the time, this convinced him that the doctrines described were 'contemptibly absurd'.[57] However, the following year Combe attended the dissection of a brain by Spurzheim, the great German populariser of phrenology, at the house of a friend. After attending a course of lectures by Spurzheim, he 'arrived at complete conviction of the truth of Phrenology'.[58] His growing interest in the subject led him to found the Edinburgh Phrenological Society in 1820. In the *Elements of Phrenology* Combe neatly summarised the essentials of his doctrine as follows: 'the brain is the material instrument by means of which the mind acts, and is acted upon; and it is a congeries of organs'.[59] He published a number of works on phrenology, including *Essays on Phrenology* (1819) and *A System of Phrenology* (1824), but it was *The Constitution of Man* published in 1828 that was to be his most influential and best-selling work.

However, as we will see below, Combe's *Constitution* was much more than just an exposition of the doctrine of the phrenological organs, which only accounts for a relatively small proportion of the work. For Combe's vision of human nature was not a static one, but a story of progress and development. One of the central tenets of the *Constitution of Man*, an essential part of what Chambers described as the 'metaphysical department of the science', was the doctrine of universal progress and the perfectibility of human nature. Combe not only found the idea of progress built into the constitution of man, he also found analogies for the progress of society in the progressive constitution of the world. In particular, he considered his views of human nature in perfect harmony with the progressive story of the history of the globe presented by contemporary geology. As he wrote in the 'introductory remarks' to the 'people's edition' of *Constitution*:

> The constitution of the world ... appears to be arranged in all its departments on the principle of slow and progressive improvement. Physical nature itself has undergone many revolutions, and apparently has constantly advanced. Geology seems to show a distinct preparation of it for successive orders of living beings, rising higher and higher in the scale of intelligence and organisation, until man appeared.[60]

In addition to the principle of universal progress, Combe also emphasised the centrality of natural law. He expressed this in a letter to Robert Chambers in December 1833, where he stated: 'The leading principles of the "Constitution of Man", to which I attach value, & which seem to myself to be original, are, the separate existence & operation of each natural law; the necessity of obeying all of them; & the evident adaptation of all to the moral & intellectual advancement of the race.'[61] The natural laws could be divided into three categories: the physical, the organic and the moral or intellectual laws. Following some of the examples given by Combe, the physical laws are those which determine that an unsound ship will sink, while a sound one will float; the organic laws determine that an individual who has a good diet and takes exercise will be healthy, while one who has a bad diet and does insufficient exercise will become ill; and the moral laws ensure that a person who 'obeys the precepts of Christianity, will enjoy within himself a fountain of *moral and intellectual happiness*'.[62] These natural laws had been instituted by the deity for the governance of the world. Constant divine intervention was therefore not necessary to realise God's purposes on earth. This idea was to be absolutely central to Chambers' argument in *Vestiges*, where he stated:

> To a reasonable mind the Divine attributes must appear, not diminished or reduced in any way, by supposing a creation by law, but infinitely exalted. It

is the narrowest of all views of the Deity, and characteristic of a humble class of intellects, to suppose him acting constantly in particular ways for particular occasions.[63]

For Combe, therefore, the secret of the progress of human society and the perfection of human nature lay in the action of secondary laws, put in place by the deity. To understand and follow these laws was the key to progress. The natural laws ensured that the provision of sustaining and stimulating conditions would bring about the progressive development of both individuals and society.

There is clearly a close analogy between the picture of human progress through the positive influence of social conditions envisaged here and the progressive development of increasingly advanced forms of animal life through the action of favourable conditions on the mother proposed by Chambers. We have seen above that natural laws played a crucial role as secondary causes, underlying the universal progress that Chambers observed in nature. For Combe too, 'every mode of action, which is said to take place according to a natural law, is inherent in the constitution of the substance of being'.[64] It can be concluded that Combe's ideas of progress through natural law had a profound influence on Chambers and are in evidence throughout *Vestiges*. Both he and many of the transformist thinkers we have examined believed passionately in the reality of universal progress and the supremacy of natural law, beliefs with their roots in late-Enlightenment optimism. In this they do have a deep philosophical kinship, even if no more direct link can be established between them. It is no coincidence that phrenology was also one of the most popular topics for debate among the medical students of the Plinian Natural History Society and the Royal Medical Society in the late 1820s and early 1830s. Combe may not have been a source of the theory of transformism proposed by Chambers, or even any of its elements, but his writings held out the prospect of continuous, law-governed progress of the kind that also made transformism such a compelling idea for Chambers and for many of his readers.

There is no denying that Combe's advocacy of universal progress provided the central inspiration for Chambers' work. The overarching emphasis on progress through natural law is clearly derived from this source. It would indeed be difficult to imagine that *Vestiges* would have been written at all without the influence of Combe's optimistic progressivism. In addition, it seems likely that Chambers also owed his emphasis on the effects of the conditions of life to Combe. Combe saw favourable conditions for people as a prerequisite for the fullest development of the individual as well as for social progress, while Chambers transferred this idea to the whole of the living world through the power of favourable conditions to prolong gestation, allowing new, more perfect forms to appear.

As regards Chambers' knowledge of the works of earlier transformists, the evidence of *Vestiges* indicates at best a limited and superficial knowledge of older transformist theories. Chambers repeated a tired and erroneous misinterpretation of Lamarck's theories, even although he seems to have been familiar with Lyell's much more fair and accurate appraisal of Lamarckism. This raises the question of why Chambers repeated the more hostile account of Lamarck's theories when he had access to the more sympathetic one provided by Lyell. Perhaps he simply wanted to differentiate his theory from Lamarck's, which clearly it has more in common with than Chambers might have wanted to admit, but it is impossible to know for certain. There is no evidence that Chambers had any knowledge of Geoffroy's transformist speculation, with which Chambers' theory in any case has much less in common. This is not to say, however, that Chambers owes nothing to Geoffroy. His work is steeped in ideas derived from the works of Edinburgh-educated medical writers who were profoundly influenced by Geoffroy and other continental comparative anatomists and embryologists. Almost every element of his theory of transformism can be found in the Edinburgh extra-mural lecturer John Fletcher's *Rudiments of Physiology*, including the idea of unity of plan, the recapitulation theory of foetal development and the crucial connection between increased length of gestation and the development of more advanced forms.

In summary, the overarching natural-theological underpinnings of Chambers' development hypothesis, in which immutable natural laws established by the deity ensure universal progress, seems to have been borrowed wholesale from Combe's *Constitution*. On the other hand, the details of his theory of 'the universal gestation of nature' can practically all be found in Fletcher's *Rudiments of Physiology*, with some important modifications based on Von Baer's embryology, almost certainly derived from Carpenter's *Principles of General and Comparative Physiology*. While it is not impossible that Chambers drew some of these ideas from other sources, it is striking that his entire theory could well have been derived from his reading of these three sources without leaving many obvious gaps. Although Chambers was broadly aware of some earlier transformist theories, notably that of Lamarck, his knowledge of their details seems to have been at best sketchy. While they may have sown the seeds of the possibility of the transmutation of species in his mind, he does not seem to have relied on them to any significant extent when developing his own theory. Although Fletcher rejected transformism himself, his book did hint at transformist interpretations of the evidence of comparative anatomy and embryology. Coupled with the concept, derived from Carpenter, that all animals diverged off from a common developmental track during foetal development, this formed the backbone of Chambers' theory of 'the great gestation of nature'. He presented these in such a way that Chambers would have had little work to do to tailor these ideas to transformist ends, in line with his belief in the principle of universal progress he shared with Combe. Fletcher therefore provides a very real link between the ferment of new ideas that supported transformist interpretations of the natural world in Edinburgh in earlier decades and Chambers' transformist magnum opus. Although he may have come to his theory of transformism in an indirect way, it would almost certainly not have been possible to construct it at all if it had not been for the flowering of interest in new ideas in medical and natural history circles in the relatively tolerant atmosphere of the Edinburgh of the 1820s and early 1830s.

## TRANSMUTATION WITHOUT PROGRESS: ROBERT KNOX AND HEWETT COTTRELL WATSON

Although little is heard of transformism in Edinburgh natural history circles between the mid-1830s and the publication of *Vestiges* in 1844, some significant individuals who left the city to establish careers elsewhere subsequently revealed themselves as believers in the transmutation of species, although they had not made such views known during their years in Edinburgh. Two figures who were active in Edinburgh natural history circles in the decade between 1825 and 1835, Robert Knox and Hewett Cottrell Watson, stand out as important advocates of transformism, although neither espoused the transmutation of species in print until much later.[65] Both Knox and Watson are also remarkable in having advocated versions of transformism while simultaneously rejecting the evidence for progress in the natural world.

We have seen in Chapter 3 that Knox was a vociferous opponent of progressivist visions of the history of life, and was deeply critical of Lamarckian transformism on those grounds. This did not mean, however, that he believed in a static model of the history of life on earth. 'That Knox held to a theory of organic descent is beyond question' was Evelleen Richards' conclusion, based on a detailed analysis of his later works.[66] She also sought to demonstrate that his theoretical views were both strikingly original and radically different from those of either Lamarck or Geoffroy. Unfortunately Knox wrote practically nothing on the subject of transformism and the relationship between species and genera during his years in

Edinburgh, but only did so much later, when he was resident in London in the 1850s. Although, of course, it is not possible to extrapolate with any certainty from Knox's theories of the 1850s back into the 1820s, it does seem worth exploring some of the evidence that can be mustered in defence of Richards' interpretation.

In Knox's later works, he presented a theory of the relationship between the species and the genus that have led to some confusion over his views on transformism. A keen fisherman, Knox gave the clearest exposition of his theory using the example of the salmon. For Knox, the young salmon was a 'generic animal', displaying the characteristics of all species of the genus. In the words of an article from *The Lancet* in 1855: 'In the young of the true salmon, I found the specific characters of all the sub-families of the genus present; that is, red spots, dark spots of several kinds, silvery scales, proportions, and a dentition identical. The young fish before me was, in fact, a generic animal, including within it the specific characters of all the species composing the natural family.'[67] The young fish then differentiate, losing some of the 'generic' characteristics and becoming members of one or another species. In an article for *The Zoologist* in the same year, he develops this idea further:

> Thus the young animal, at a certain stage of its growth, is the type not of the species to which it belongs by hereditary descent, but represents the generic type, transcendental, and requiring for its full development or embodiment in all its material, that is, specific forms, countless millions of years; for as the young, that is, the generic animal, includes many species, perhaps all which the natural family can assume in time and space, so as species die out, others appear, new to the world as species, but not generically.[68]

It is with this definition of a species in mind that we must read his pronouncement that 'I adhere to the same view – namely the inconvertibility of species into each other by any physical laws now in operation'.[69] Under Knox's definition of a species, this need not contradict his assertion elsewhere that 'I believe all animals to be descended from primitive forms of life'.[70] He even goes as far as to say that '[i]n time there is probably no such thing as species'.[71] For Knox, the fundamental unit of nature was not the species, but the genus, so when he speaks of the impossibility of the transmutation of *species*, this does not contradict his open avowals of transformism elsewhere.

In Knox's theory the process of differentiation of the generic animal to give the species is distinct from, but preceded by, its ascent through all consecutive degrees of organic complexity during foetal development, as proposed by Serres and Geoffroy: 'from the moment of conception or of independence, that living point, that embryo, passes through a succession of *forms*, shadowing forth the organic world as it now exists, from the highest to the lowest; shadowing forth the organic world as it existed from the dawn of creation to the present day – that is proved by geology.'[72] It is worth noting here that Knox assumed that the succession of forms in order of complexity would also recapitulate the history of life through geological time as evidenced by the fossil record. This did not mean, however, that he saw the history of life as in any sense progressive; indeed, Knox poured scorn on humanity's presumption that they were the pinnacle of creation, towards which the history of life had been leading, pointing out that the fishes that inhabited the primordial oceans would, had they been able to speak, probably have said the same of themselves.[73]

Another Edinburgh graduate whose ideas on the relationship between varieties, species and genera bear comparison with those of Knox, as well as harking back to the ideas of Linnaeus on hybridisation, was the botanist Hewett Cottrell Watson. Watson studied medicine at the University of Edinburgh between 1828 and 1832, but left without graduating. We know from his own account that he attended Robert Jameson's natural history course during his student years.[74] He was not, however, a member of the Plinian Society and his name does not appear in any of the Society's records. Nonetheless, he did share interests with many members of the Society. For example, while in Edinburgh, Watson got to know the leading phrenologist George Combe and developed a deep interest in phrenology. After leaving Edinburgh he maintained close links with the city both through his friendship with Combe and his continuing association with Edinburgh botanical circles. In 1836 he helped to found the Botanical Society of Edinburgh.

In his monograph on Watson, Frank N. Egerton claimed Watson as a transformist, although only after he left Edinburgh.[75] In his foreword to Egerton's book, David L. Hull states that 'Watson had accepted the transformation of species from at least 1834'.[76] I would argue that Watson's surviving correspondence and published work suggest that he was certainly never a transformist of the same stamp as, for example, Robert Grant. Like Knox, he was highly sceptical, if not actively hostile, to the idea of progress in the natural world. In an outline of the 'Progress of the Earth's Changes' he sent to George Combe in December 1836, he wrote of the fossil evidence for the history of life that 'I think the evidence shows oscillation to & fro, without any onward or backward course in continuity'.[77] His transformism, then, if it can be described as such, was of an unusual uniformitarian variety. This did not mean, however, that he believed that new species did not come into existence over time. It is to his ideas on the origins of new species and the relationship between varieties, species and genera that I now turn.

In his *Outlines of the Geographical Distribution of British Plants*, published in Edinburgh in 1832, Watson discussed at some length the various possible opinions on the origins of species of plant in the following passage:

> Investigations concerning the original creation of plants, in the present state of human knowledge, might be deemed by many at best an idle waste of time; and even inquiries into the means by which some occupy their present situations, except in some, and these comparatively but few particular instances, may truly seem a speculation not much more profitable in itself, or likely of ultimate success. This inquiry, nevertheless, has occupied the attention of several excellent botanists, and has lead [sic] to considerable diversity of opinion amongst them; – one party imagining all plants to have originated in some central point from which they have been gradually spread over the earth's surface; others conceiving that several of such centres must have existed; and a third party believing species for the most part to have originated where they now appear as natural and untransported products of the soil and climate. Some again suppose, that at first only *genera* existed, *species* arising from generic admixture; while others maintain that all vegetable forms are modifications of each other, or the result of certain concurrence of molecules dispersed through matter; hence liable to be produced in any situation where the necessary conditions of their existence occur.[78]

First Watson issued a caveat warning of the doubtful value of speculation, before reviewing the various explanations for the distribution of plants that were then current. He then turned his attention to the various theories offering an explanation for the appearance of new species. He considered three possibilities: first, Linnaeus' model of the formation of new species through hybridisation; second, the transmutation of existing species; and third, the model elaborated by Buffon in his *Époques de la nature*, in which new species arise spontaneously from 'organic molecules' wherever conditions are suitable for them. He was therefore well aware of the vexed question of the origin of species and of some of the answers proposed to it, although at this point he does not seem to have favoured the transmutation of species over the other possibilities.

Like Buffon, Lamarck, Knox and Cheek, Watson seems to have doubted that species had any real existence in nature, instead believing them to be merely a product of the human need to reduce the world to order. Watson came to believe that species were fluid entities that were in a constant state of flux. As he wrote in a letter to Nathaniel Winch in October 1834, '*Species* in any sense or degree I look on as human divisions, not as the creations of nature. The changes, provided by geologic evidence, to have occurred in organic forms, and those now effecting by climate, elevation, cross-breeding, &c. &c. strongly discountenance the idea of absolute and permanent distinctions.'[79] In a key work entitled *An Examination of Mr. Scott's Attack upon Mr. Combe's 'Constitution of Man'*, in which Watson set out to defend the ideas of his friend and fellow phrenologist George Combe from the attacks of his Evangelical critic, William Scott, he suggested that there were no fixed boundaries between species:

> we find varieties [of plants] produced, and regularly continued by descent, having greater differences between themselves, than are seen between other races generally supposed to be distinct species. So much do our gardens now abound with intermediate varieties or transition-species, so generally is one kind run into another, that the united skills of all the botanists in the world would fail to distinguish them.[80]

He backed up his argument with evidence drawn from the production of breeds of domestic animals and plants, although, at least as regards higher animals, he was less than sure that the evidence was conclusive, but only admitted that 'it "tends" to show a possibility of change and progression'.[81] In response to Scott's use of Cuvier's argument that historical records and the discoveries of archaeology show no discernible change in species, Watson gives the same reply as Lamarck and Geoffroy, pointing out that the two or three thousand years quoted by Cuvier represents 'a space of time which shrinks to a mere point, if compared with the eras of geologists'.[82] Watson's conception of the history of life, then, is one in which there is constant flux and change, but no clear sign of progressive development.

Both Knox and Watson were figures whose careers and opinions were shaped by their experiences in Edinburgh natural history circles in the 1820s and 1830s. In Knox's case, his studies in Paris and his position as an extra-mural anatomy lecturer also put him in a prime position to act as a conduit for the transfer of the latest theories from France to Edinburgh. Although we lack hard evidence that either was a convinced transformist while resident in Edinburgh, they both certainly became so later in their careers, in Watson's case within a year or two of terminating his studies at the University. In the cases of both men their apparent espousal of transformism combined with a denial of universal progress presents a puzzle. Knox was something of a showman in his teaching as in his writing. Both advocating the transmutation of species and denying the reality of progress would be highly controversial stances in the 1840s and 1850s. It is conceivable that Knox, whatever his personal opinions on the matter, was simply playing with these ideas in his published work with the aim of stirring up controversy, perhaps with an eye on promoting the sales of his books. He certainly lived to a considerable extent by his pen and by giving public lectures after his move to London in the 1840s, and such publicity would have done him no harm, given his already decidedly chequered reputation. This would make it almost impossible to determine what his true opinions were on the subject. In any case, he was certainly engaging with transformist theories in his published work at this time. For the more self-effacing Watson this is a less likely explanation, as he would seem to have had little to gain and much to lose from courting notoriety in this way. However, unlike Knox, Watson never made any sweeping statements claiming that all animals were 'descended from primitive forms of life', and his vision of new species arising as variations around a mean rather than appearing in a progressive sequence did not reveal the inconsistencies that are so evident in Knox's writings. In any case, their shared conviction that the history of life on earth did not display an upward trajectory was to find few followers in later decades.

## THE LEGACY OF DARWIN'S EDINBURGH YEARS

Charles Darwin, the man whose theory of evolution was finally to make the concept of the transmutation of species acceptable to a new generation of natural historians, was a student at the University of Edinburgh at just the time transformism and other related ideas were being widely debated there. Yet his account of the development of his own theory in his autobiographical and other writings has little to say about his experiences in Edinburgh. In his 'Recollections', published in 1876, we find his only admission that he had heard evolutionary ideas spoken of during his time at the University of Edinburgh. Here he claimed that on one occasion Robert Grant had praised Lamarck's ideas in his presence, but that he had 'listened in silent astonishment, and as far as I can judge, without any effect on my mind'.[83] It is clear that Darwin was suggesting that any evolutionary theorising he was exposed to in Edinburgh had no effect on his later development of his own theory. Should we take him at his word on this, or might he have had good reasons to deny the very real influence of earlier transformists on the genesis of his own theory?

There are good reasons why Darwin may have wanted to present his theory as the fruit of long years of patient data collection and downplay any influence of earlier transformist thinkers. He was profoundly anxious not to be dismissed as another armchair theoriser, an image, rightly or wrongly, that had become associated with earlier transformist thinkers such as Lamarck and Darwin's own grandfather Erasmus Darwin. He was equally concerned not to be seen as another bungling dilettante, a perception that had allowed the intellectual elite of mid-nineteenth-century science to dismiss the work of the anonymous author of *Vestiges* out of hand. In the 1830s there was growing unease about the status of the sciences and increasing pressure for science to become professionalised from commentators such as David Brewster and Charles Babbage. Babbage's *Decline of Science in England*, published in 1830, bemoaned the fact that 'The pursuit of science does not, in England, constitute a distinct profession, as it does in many countries', and called vociferously for the increased professionalisation of science and its practitioners.[84] In such a climate it was vitally important that Darwin prove his credentials as more than a gentleman of leisure who dabbled in natural history, especially since he neither wished nor needed to earn his living from science, hold a university chair or engage extensively as the officer of any society.

We know that Darwin read and was deeply influenced by John Herschel's *Preliminary Discourse on the Study of Natural Philosophy* (1830) not long after he sat his BA examinations at Cambridge in 1831.[85] This work has been described as a 'formative text on "scientific method"' by John V. Pickstone and provided a handbook for the correct conduct of scientific investigations.[86] Contemporaries regarded it as the greatest work on scientific method since the writings of Francis Bacon, whose bust appeared on its half-title page.[87] Herschel wrote: '[w]henever, therefore, we would either analyse a phenomenon into simpler ones, or ascertain what is the course or law of nature under any proposed contingency, the first step is to accumulate a sufficient quantity of well ascertained facts, or recorded instances, bearing on the point in question.'[88] The role of theory was to be tightly circumscribed: 'The liberty of speculation which we possess in the domain of theory is not like the wild licence of the slave broke loose from his fetters, but rather like that of the freeman who has learned the lessons of self-restraint in the school of just subordination.'[89] Darwin seems to have deeply imbibed the lessons of Herschel's book that data gathering was the primary activity of the natural philosopher and that theories must always emerge from the facts rather than determining the direction of research. In later years, Darwin suggested that his theory of evolution had sprung, as Herschel suggested good scientific theories should, from the empirical evidence he had collected in his travels and researches. In his autobiography, written decades after the period of his life he was describing, he said of his research methods that he had 'worked on true Baconian principles, and without any theory collected facts on a wholesale scale'.[90] The years he spent between 1846 and 1854 establishing himself as a world expert on the biology of barnacles can be seen as part of his 'self-fashioning', as a natural historian to be taken seriously rather than as a mere dabbler. The image he projected of a model Baconian philosopher was to play a significant role in making his theory acceptable to his elite scientific peers, but it could be argued that it also obscured the roots of some important elements of his theory.

Frank Sulloway has provided us with an excellent case study that throws into question the extent to which Darwin's early ideas on evolution did indeed flow directly from the raw data he collected.[91] Sulloway has shown that Darwin's observations on the Galapagos finches did not play the essential role in the genesis of his thought that Darwin later suggested and popular accounts of his theory have insisted ever since. Indeed, when he collected the specimens he failed to label them accurately with the names of the islands where he collected them. This information had largely to be gleaned much later from the better documented collections made by some of Darwin's shipmates. It was John Gould, the eminent ornithologist, who helped Darwin with the classification of the birds he collected, and who realised that the Galapagos finches were all very closely related and did not belong to the six or seven genera that Darwin had supposed.[92] Sulloway concludes from this that 'Darwin's finches do not appear to have inspired his earliest theoretical views on evolution, even after he finally became an evolutionist in 1837; rather it was his evolutionary views that allowed him, retrospectively, to understand the complex case of the finches'.[93] The impression that Darwin interpreted the finches in the light of his developing theory rather than forming the basis of it is reinforced by the fact that they are nowhere mentioned in the four notebooks he filled with notes on transformism between 1837 and 1839. It was only with his 'Essay of 1844' that Darwin developed a compelling picture of how the evolution of the finches could have taken place. If the earliest version of Darwin's theory of evolution did not always emerge 'on true Baconian principles' from the data he had collected on his voyage, where did it come from?

The genesis of Charles Darwin's theory of evolution by means of natural selection has been the subject of endless debate on the part of the scholars of the 'Darwin industry', and it is not my intention to retell a tale so frequently told before by so many distinguished historians of science. Instead, I will limit myself to a discussion of the extent to which the development of Darwin's ideas leading up to the publication of the *Origin of Species* can be seen as continuing the preoccupations of the Edinburgh transformists of the 1820s, among whom Darwin spent the years 1825 to 1827. In order to do this I will consider not so much the Darwin whose theory was finally published in the *Origin of Species* in 1859, but the Darwin of the *Beagle* years and the Darwin who recorded his developing ideas on the transmutation of species in a series of notebooks after his return to England, leading to the 'Essay of 1844' in which he gave a résumé of his ideas to date. To what extent did the researches of Darwin between leaving Edinburgh and the publication of the *Origin of Species* reflect the preoccupations that animated the natural history circles around Robert Jameson and the Plinian Natural History Society in the 1820s?

In answering this question I do not aim to determine whether or not Darwin did in fact imbibe certain doctrines while at Edinburgh, which is clearly not possible on the basis of the surviving evidence. Rather I intend to establish which ideas that appear in his later work would seem to have grown out of the debates on the nature of species and their origins that were taking place in the Edinburgh of the 1820s. It is, of course, impossible to prove that Darwin did not absorb these ideas from other sources at a later date. And indeed he continued to be surrounded by former Edinburgh University students in metropolitan natural history circles while he was marshalling his evolutionary ideas in the notebooks and essays he wrote in the seven or so years after his return from the voyage of the *Beagle*; William Benjamin Carpenter, Richard Owen and Hugh Falconer, for example, all studied at Edinburgh in the 1820s and 1830s and could have acted as later conduits for ideas and influences. Nonetheless, the ultimate source of these ideas is the same.

Phillip R. Sloan has written in some detail about the 'ever-present biological concerns that can be followed in Darwin's thought in an unbroken line from his earliest Edinburgh days through the Cambridge and Beagle years into the mature writings'.[94] We know from his own account of that time that Darwin spent a considerable amount of time with Robert Grant in his second year in Edinburgh.[95] Like Darwin, Grant was a familiar presence at the meetings of the Plinian Society in the mid-1820s. Grant's involvement with the Society was slightly anomalous, as the Plinian was a student society and Grant was not only not a student but was considerably older than most of the other members. He seems to have acted as a mentor figure to many of the student members, including the young Darwin. Darwin became very involved with Grant's researches and presented a paper giving some of the fruits of his work to the Plinian Society on 27 March 1827 in which:

Mr Darwin communicated to the Society the discoveries that he had made –

1. That the ova of the Flustra possess organs of motion.

2. That the small black globular body hitherto mistaken for the young Fucus lorins, is in reality the ovum of the Pontobella muricata.[96]

But three days before Darwin read his paper to the Plinian Society, Grant read a paper to the Wernerian Natural History Society in which, along with giving an account of his own researches, he included a description of Darwin's observations regarding Flustra and *Pontobella* without any acknowledgement of the younger man.[97] Even though Grant acknowledged his 'zealous young friend Mr Charles Darwin' when some of this material was subsequently published in David Brewster's *Edinburgh Journal of Science*, Darwin seems to have been deeply offended by Grant's ungracious appropriation of his work.[98] Although his relations with Grant appear to have cooled significantly as a result of this incident even before Darwin left Edinburgh, his largely unacknowledged influence on the young naturalist seems to have been more durable than their friendship.

Grant was not the only known transformist whom Darwin would have met at the Plinian Society. Henry H. Cheek was also a regular fixture at the Society's meetings, and he and Darwin were both present at no less than sixteen meetings between December 1826 and April 1827.[99] As attendance at these meetings could be as low as twelve members at this time, it is highly unlikely they were not acquainted. Both Darwin and Cheek were also members of the Royal Medical Society, Cheek joining on 3 November 1826, while Darwin joined on 17 November of the same year.[100] Evidence that Darwin and Cheek moved in the same student circles is also to be found in Mary Clementina Hibbert Ware's biography of her father in law, Samuel Hibbert Ware (1782–1848), a keen amateur natural historian resident at that time in Edinburgh, in which the author comments that Darwin frequently went on natural history trips with William Francis Ainsworth, Cheek's friend and co-editor of the *Edinburgh Journal of Natural and Geographical Science*.[101] In the circumstances it seems almost impossible that Darwin would not have been acquainted with Cheek and have been exposed to his views on the transmutation of species.

Darwin's enthusiasm for the biology of marine invertebrates he shared with Grant did not cease when he left Edinburgh. At Cambridge, where he studied from 1827 to 1831, his attention was drawn more towards botany and geology, but the opportunity to take a trip around the world as ship's naturalist on the famous 'Voyage of the *Beagle*' in 1831–6 allowed him to revive the interest in marine invertebrates that Grant had inspired.[102] The importance of this 'Grantian' invertebrate programme for Darwin is evident in a letter written to his old Cambridge mentor John Stevens Henslow (1796–1861), which he wrote from Rio de Janeiro in May 1832. In this letter Darwin insisted that 'Geology & the invertebrate animals will be my chief object of pursuit throughout the whole voyage'.[103] As Sloan has noted, 'Darwin's research into invertebrate zoology had begun as a continuation of the researches of Robert Grant, and they retained many marks of this Grantian heritage through the *Beagle* period'.[104] It has been calculated that in the diary of his activities that he kept on board the *Beagle*, approximately 66 per cent of the total volume of discussion is of invertebrate zoology, greatly exceeding the space devoted to vertebrate animals.[105]

Jonathan Hodge has convincingly argued that the enthusiasm for studying generation, or reproduction in modern terms, among the most simple invertebrate animals that Darwin picked up from Grant was to turn him into a 'lifelong generation theorist'.[106] When Darwin turned to transformism in the notebook he began filling with observations and speculations on his return from his epic journey on the *Beagle* these ideas on generation, and in particular the role that generation played in producing variation in living things, were to form one of the cornerstones of his emerging theory. Generation, or the mechanism by which animals and plants reproduced themselves, was, and was to remain, central to Darwin's theories; after his development of the concept of natural selection it was the source of the variations in living things on which selection could act. In one of the first few pages of 'Notebook B', the first of the notebooks that Darwin started on transformism after the return from his voyage, he wrote: 'We see ~~living beings~~ the young of living beings, become permanently changed or subject to variety, according to circumstances, – seeds of plants sown in rich soil, many kinds, are produced, though new individuals produced by buds are constant, hence we see generation here seems a means to vary, or adaptation.'[107] Darwin, with no clear idea of a mechanism of inheritance, postulated that environmental factors acted on living things during the early stages of their development (but not on adult organisms) causing them to vary, and that these variations would then be heritable.

Dov Ospovat has argued that Darwin's understanding of the nature of variation changed dramatically after reading Malthus towards the end of 1838. Before this date, Darwin believed that 'the generative system, in response to changing external conditions, produces variations that are adapted to the altered circumstances'.[108] There was therefore some 'zoological law', which ensured that the variations that occurred would be adaptive. These changes would occur and become heritable only during the early development of the organism, not in the adult form. The belief that changing the conditions of life would automatically generate variations in living things is familiar from the writings of Geoffroy, whose 'experimental transformism' aimed to induce variation in chickens by altering the conditions in which the eggs were incubated. Geoffroy only published a detailed exposition of his theory in 1828, but Darwin may well have known about his researches directly or indirectly through Grant or Robert Knox, who maintained close contact with developments in French natural history circles in general and with Geoffroy in particular.[109] Geoffroy had first published an explicitly transformist paper in 1825, which Darwin might certainly have been aware of during his time in Edinburgh.[110] The records of the Plinian Society suggest that Edinburgh medical and natural history circles were abuzz with talk of Geoffroy's theories of unity of type at around the time Darwin was a student there, so it is unlikely that Darwin remained unacquainted with his ideas.

When Darwin later came to formulate his key concept of natural selection after reading Thomas Malthus' *Essay on the Principles of Population* in late 1838, this Geoffroyan mechanism would generate the variations on which natural selection could act. By this time, Darwin seems to have come to the conclusion that variations were not necessarily adaptive. Instead, he came to believe that 'every animal produces in course of ages ten thousand varieties (influenced itself perhaps by circumstances) & those alone preserved which are well adapted'.[111] Darwin was to maintain this essential position throughout his life. In the sketch of his theory he wrote in 1844, he was even more explicit about how the action of natural selection on random variation caused by unfavourable conditions would cause evolutionary change, while in the absence of selective pressures, intra-species variation would lead nowhere due to free interbreeding between variant and normal forms:

> Hence almost every part of the body would tend to vary from the typical form in slight degrees, and in no determinate way, and therefore *without selection* the free crossing of these small variations (together with the tendency to reversion to the original form) would constantly be counteracting this unsettling effect of the extraneous conditions on the reproductive system.[112]

Darwin's entire model of the action of variation and selective pressure exerted by the organism's conditions of life was, of course, very close to Geoffroy's opinion published a few years earlier in 1831 that '[i]f those modifications lead to harmful effects, the animals which undergo them will cease to exist, to be replaced by others, with forms modified to suit the new circumstances'.[113] We do not know if Darwin ever read these words, but the parallels with his own views at around the time he was formulating the theory of natural selection are striking. We know that Geoffroy never developed a fully worked out theory of evolution by means of natural selection, but he was certainly thinking along these lines in the late 1820s and early 1830s. Darwin later came to see the pressure of competition both with other members of the species and with other species as more important than the pressure of changing conditions. However, in the picture of evolution driven by random variation and directed by natural selection, we have the essence of Darwin's argument in *The Origin of Species* and beyond.

Darwin's notebooks are also replete with references to Geoffroy's older colleague at the Jardin des Plantes, Jean-Baptiste Lamarck. Darwin was often dismissive of what he described as 'Lamarck's "willing" doctrine', referring to the popular misconception that Lamarck had suggested that animals literally 'willed' themselves to evolve, which Darwin at times seems to have shared.[114] Elsewhere, however, in Notebook C for example, he admitted that Lamarck 'had few clear facts, but so bold & many such profound judgement that he foreseeing consequence was endowed with what may be called the prophetic spirit in science. the [sic] highest endowment of lofty genius'.[115] This is quite contrary to the impression sometimes given that Darwin was completely dismissive of Lamarck's work.

Jonathan Hodge has convincingly argued that at the time Darwin started recording his ideas in 'Notebook B' in July 1837, 'one has, more than anything else, to look to Darwin's relations at this time with four sources of precedential instruction and inspiration: Lyell, Lamarck, Grant and his own grandfather, Erasmus Darwin'.[116] Darwin encountered Lyell's theories after his time in Edinburgh, but we know Grant was immersed in the ideas of both Lamarck and Erasmus Darwin in the 1820s when Darwin was spending a significant amount of time in his company. It seems extremely likely that Darwin would have encountered the ideas of Lamarck for the first time during his collecting trips with Grant, and Hodge has also suggested that Grant probably encouraged the young naturalist to study the works of his grandfather, especially his *Zoonomia*. Hodge indeed convincingly argues that first Grant and later Lyell were by far the two most important influences on the development of Darwin's theory.

There is a distinct continuity between the ideas of the Edinburgh transformists and the tentative theories adopted by Darwin when he turned his mind to the transmutation of species on his return to England from the voyage of the *Beagle*. Of course, he went on to take these ideas off in radically new directions and elaborate them to a much greater degree over the following two decades than was possible for any of the Edinburgh transformists. Grant, Jameson, Cheek and their Edinburgh contemporaries did not have the leisure and the opportunities available to Darwin, who was able to devote himself almost entirely to working on his theory for much of his life. Darwin wrote the following with becoming humility in Notebook D in late 1838: 'Seeing what Von Buch (Humboldt) G. St. Hilaire, & Lamarck have written I pretend to no originality of idea – (though I arrived at them quite independently & have used them since) the line of proof &

reducing facts to law only merit if merit there be in following work.'[117] However, knowing what we now know about the prevalence of transformist ideas in Edinburgh in the later 1820s, it is implausible that he came upon the ideas he shared with Geoffroy and Lamarck quite as independently as he and many subsequent historians have suggested. Dov Ospovat ended his exemplary study of *The Development of Darwin's Theory* by concluding that 'Darwin's theory represents not so much the result of an interaction between the creative scientist and nature as between the scientist and socially constructed conceptions of nature'.[118] I would argue that a significant proportion of the conceptions of nature that formed the underpinnings of his theory of evolution were present in the Edinburgh of the late 1820s, being discussed and debated by his friends, teachers and fellow students. Even if his mind was closed at the time to some of the more daring of these ideas, as he subsequently claimed, his later writings are shot through with echoes of the debates and opinions of the Edinburgh transformists of the 1820s.

By early 1834 Robert Grant was in London, Henry Cheek was dead and the generation of medical students that had swelled the ranks of the Plinian Society in the late 1820s were scattered across the British Empire and beyond. Robert Jameson was to continue as professor of natural history at the University of Edinburgh until 1854, but there is no record of any continuing interest in transformism on his part, and the stream of transformist articles in the *Edinburgh New Philosophical Journal* had dried up by the mid-1830s. Although absence of evidence is not evidence of absence, it seems that the spirit of radical enquiry into the mutability of species that had animated Edinburgh natural history circles in the 1820s had departed. It may, perhaps, have been hurried on its way by Evangelical hostility and the rise of Cuverian catastrophism to the status of orthodoxy in geology.

Nonetheless, the ferment of ideas that had gripped natural history circles in Edinburgh in the decade from 1823 to 1833 was not altogether without a legacy. A significant number of those who were to espouse transformist views in the decades before the publication of the *Origin of Species* were present in Edinburgh in those years. Figures who had moved in the natural history circles around Robert Jameson continued either to hold to the beliefs they established at that time, as did Robert Grant, or to mine the rich variety of ideas that emerged from this source to formulate their own original theories about the origins and transmutation of species, as did Knox and Watson. Chambers did his theorising at one remove from the source of his ideas, but still drew on many of the key theories current in transformist circles in Edinburgh via the writings of Fletcher, which provided a neat digest of the theories of transcendental anatomy that had also fuelled earlier transformist speculation. From these pre-formed pieces he constructed his own distinctive transformist theory, which owed much to the intellectual climate that had nurtured the previous generation of rather different transformist thinkers, of whom he himself was in all likelihood largely unaware.

And finally we come to the question of the evident continuities between the theories of the Edinburgh transformists and the later writings of Charles Darwin. It seems from the evidence presented by Sloan and Hodge that there was a high level of continuity between Grant's invertebrate programme and Darwin's research on the *Beagle*. The notebooks he kept when he turned his mind to the question of transformism after his return to England are also replete with references to the theories of Lamarck and Geoffroy, whose ideas inspired the Edinburgh transformists. The speculations he based on these sources are absolutely in the same vein as those of Grant and Cheek a decade earlier, although his reading of Malthus ultimately carried him off in a direction that was to lead him to natural selection after 1838.[119] His determination to present his theory as a model of Baconian induction made it imperative for him to deny any connection between his work and that of earlier speculative thinkers, and formed a crucial element in his 'self-fashioning' as a Victorian man of science. This was necessary insurance against sharing the fate of Chambers, dismissed as an ignorant dabbler by most of the scientific establishment in the 1840s. Nonetheless, a study of the evidence for the genesis of his theory of evolution clearly shows that the ideas that inspired the Edinburgh transformists in the 1820s were at the heart of his theory from its very beginning.

## NOTES

1. Secord, *Victorian Sensation*, p.131.
2. Buckland, *Geology and Mineralogy*, vol.1, p.130. For excellent accounts of the Bridgwater Treatises and their reception, see Topham, 'Science and popular education in the 1830s' and Topham, 'Beyond the "common context"'.
3. Sedgwick, *Discourse on the Studies of the University*, pp.25–6.
4. Ibid., p.30.
5. Cuvier, *Essay on the Theory of the Earth*, p.14. The original French text reads: 'Il y a donc eu dans la nature une succession de variations que ont été occasionnées par celles du liquide dans lequel les animaux vivaient ou que du moins leur ont correspondu ; et ces variations ont conduit par degrés les classes des animaux aquatiques à leur état actuel' (Cuvier, *Révolutions de la surface du globe*, p.14).

6. Miller, *Old Red Sandstone*, p.102. Miller's use of metaphors from visual culture is brilliantly explored in O'Connor, *The Earth on Show*, pp. 391–432.
7. [Brewster], Review of [Chambers'] *Vestiges*, p.474.
8. Miller, *Old Red Sandstone*, pp.44–5.
9. Duncan, 'Memoir of Lamarck', p.37.
10. Ibid., pp.17–18.
11. Ibid., p.36.
12. Cuvier, 'Memoir of M. de Lamarck', pp.1–22.
13. Duncan, 'Memoir of Lamarck', p.63.
14. Ibid., p.45.
15. See, for example, Brewster, Review of [Chambers'] *Vestiges*, p.374; and Miller, *Foot-Prints of the Creator*, p.15.
16. Fleming, 'Natural science', in Fleming, *Inauguration of the New College of the Free Church*, p.218.
17. Quoted in Duns, 'Memoir', in Fleming, *Lithology of Edinburgh*, p.xl.
18. Secord, *Victorian Sensation*, p.130.
19. The Chambers' publishing business has been explored in depth in Fyfe, *Steam-powered Knowledge*.
20. Pringle Nichol, in his popular *Views of the Architecture of the Heavens*. For an account of the nebular hypothesis and contemporary reactions to it, see Schaffer, 'The nebular hypothesis and the science of progress'.
21. [Chambers], *Vestiges*, p.164.
22. Ibid., p.213.
23. Ibid., pp.218–19.
24. Secord, 'Beyond the veil', pp.179–80.
25. [Chambers], *Vestiges*, p.230.
26. Ibid., p.231.
27. [Chambers], 'Popular information on science: Transmutation of species', p.274. The quote comes from Lyell, *Principles of Geology*, 3rd edn, vol.2, pp.402–3.
28. Cuvier, 'Memoir of M. de Lamarck', p.14.
29. See, for example, Lyell, *Principles of Geology*, 3rd edn, vol.2, pp.331–2.
30. In Lyell, *Principles of Geology*, 3rd edn, vol.2, p.333 and [Chambers], *Vestiges*, pp.230–1.
31. [Chambers], 'Natural history: Animals with a backbone', p.338.
32. Ibid., p.138.
33. [Chambers], 'The life and poetry of Darwin', p.251.
34. [Chambers], 'Natural history: Monkeys, apes, and orang-outangs', p.362.
35. [Chambers], 'Popular information on science: Third ages of animal life', pp.298–9.
36. Secord, 'Beyond the veil', p.175.
37. [Chambers], *Vestiges*, p.182.
38. Fletcher, *Rudiments of Physiology*, p.36.

39. [Chambers], *Vestiges*, pp.222–3.
40. Ibid., p.198.
41. Ibid., pp.200–1.
42. Lord, *Popular Physiology*, p.296.
43. [Chambers], *Vestiges*, p.224. The quote comes from Fletcher, *Rudiments of Physiology*, pp.60–1.
44. Fletcher, *Rudiments of Physiology*, pp.36–7.
45. [Chambers], *Vestiges*, p.223.
46. Fletcher, *Rudiments of Physiology*, p.61.
47. Ibid., p.16.
48. Ospovat, 'Influence of Karl Ernst von Baer's embryology', pp. 8–10.
49. See, for example, Chambers to Ireland, 29 March 1845, ff.168 verso–169 recto; and Chambers to Ireland [day and month unknown] 1847, f.26 recto.
50. Ibid. p.13. For the original of the diagram see Carpenter, *Principles of General and Comparative Physiology*, p. 197.
51. Carpenter, *Principles of General and Comparative Physiology*, pp.562–3
52. Chambers to Combe, 14 December 1833, f.47.
53. [Chambers], 'Is ignorance bliss?', pp.385–6.
54. Combe to Chambers, 13 December 1833, f.64.
55. Chambers to Combe, 25 November 1835, f.140.
56. Combe to Chambers, 26 November 1835, f.423.
57. Combe, *System of Phrenology*, p.iii.
58. Ibid., p.iv.
59. Combe, *Elements of Phrenology*, p.23.
60. Combe, *Constitution of Man*, p.2.
61. Combe to Chambers, 15 December 1835, f.13.
62. Combe, *Constitution of Man*, p.6.
63. [Chambers], *Vestiges*, p.156.
64. Ibid., p.8.
65. The most important work advocating Knox as a transformist is probably Desmond, *Politics of Evolution* while Frank N. Egerton's work *Hewett Cottrell Watson* musters the evidence that Watson was a transformist.
66. Richards, 'The "moral anatomy" of Robert Knox', p.399.
67. Knox, 'Introduction to Enquiries into the philosophy of zoology', p.627.
68. Knox, 'Philosophy of zoology. Part I', pp.4789–90.
69. Knox, 'Introduction to Enquiries into the philosophy of zoology', p.626.
70. Knox, *Great Artists and Great Anatomists*, p.109.
71. Knox, *Races of Men*, p.36.
72. Ibid., p.421.
73. Knox, 'Introduction to Enquiries into the philosophy of zoology', p.369.
74. Watson, *Remarks on the Geographical Distribution of British Plants*, p.28.
75. Egerton, *Hewett Cottrell Watson*, p.148,
76. Hull, 'Foreword', in Egerton, *Hewett Cottrell Watson*, p.xix.
77. Watson to Combe, 14 December 1836, f.161 recto.

78. Watson, *Outlines of the Geographical Distribution of British Plants*, pp.2–3.
79. Watson to Winch, 7 October 1834, quoted in Egerton, *Hewett Cottrell Watson*, p.148.
80. Watson, *An Examination of Mr. Scott's Attack*, p.27.
81. Ibid., p.26.
82. Ibid., p.25.
83. Darwin, 'Recollections of the development of my mind and character', p.24.
84. Babbage, *Reflections on the Decline of Science in England*, pp.10–11.
85. Sloan, 'Making of a philosophical naturalist', in Hodge and Radick (eds), *Cambridge Companion to Darwin*, p.27.
86. Pickstone, *Ways of Knowing*, p.148. It is also discussed at some length in Chapter 2 of Secord, *Visions of Science*.
87. Secord, *Visions of Science*, p.80.
88. Herschel, *Preliminary Discourse*, pp.118–19.
89. Ibid., pp.190–1.
90. Darwin, 'My several publications', in Darwin, *Autobiographies*, p.72.
91. Sulloway 'Darwin and his finches', pp. 1–53.
92. Ibid. p.10.
93. Ibid., p.32.
94. Sloan, 'Darwin, vital matter, and the transformism of species', p.372.
95. Darwin, '1876 – Recollections of the development of my mind and character', p.24.
96. Minutes of the Plinian Society, vol.1, f.57.
97. For a full account of the incident, see Browne, *Charles Darwin: Voyaging*, pp.86–7.
98. Grant, 'Notice regarding the ova of the *Pontobdella muricata*', p.161.
99. Minutes of the Plinian Society, vol.1.
100. Royal Medical Society, List of Members, p.44.
101. Hibbert Ware, *Life and Correspondence of the Late Samuel Hibbert Ware*, p.314.
102. There has been some debate on Darwin's exact status on board the *Beagle*. Jacob W. Gruber, for example, has suggested that Darwin was in fact only a 'gentleman companion' to Captain Robert Fitzroy, while Robert McCormick, the ship's surgeon, was in fact the official naturalist. (See Gruber, 'Who was the *Beagle*'s naturalist?') Here I have followed John van Wyhe's convincing recent argument that Darwin was in fact the ship's naturalist. (See Van Wyhe, 'My appointment received the sanction of the Admiralty'.)
103. Darwin to Henslow, 18 May 1832, in *Correspondence of Charles Darwin*, vol.1, p.237.
104. Sloan, 'Darwin's invertebrate program', p.111.
105. Ibid., p.89. Peter J. Bowler has suggested that Darwin deliberately downplayed the importance of his invertebrate researches in the published account of the voyage. See Bowler, *Darwin: The Man and his Influence*, p.59.
106. Hodge, 'Darwin as a lifelong generation theorist', pp.207–43.

107. Darwin, Notebook B, p.3.
108. Ospovat, *The Development of Darwin's Theory*, p.58.
109. Geoffroy and Serres, 'Rapport fait à l'Académie Royale des Sciences sur une mémoire de M. Roulin', pp.201–8.
110. Geoffroy, 'Recherches sur l'organisation des gavials', pp.97–155.
111. Darwin, Notebook B, p.90.
112. Darwin, 'Essay of 1844', p.85.
113. Geoffroy, 'Quatrième mémoire', p.79.
114. Darwin, Notebook B, p.216.
115. Darwin, Notebook C, p.119.
116. Hodge, 'The notebook programme and projects of Darwin's London years', in Hodge and Radick (eds), *Cambridge Companion to Darwin*, p.49.
117. Darwin, Notebook D, p.69.
118. Ospovat, *The Development of Darwin's Theory*, p.229.
119. It is worth noting here that it has been plausibly suggested by John S. Warren that Darwin also drew the ideas of natural selection from the writings of earlier natural historians, and in particular those of the Scots-American physician William Charles Wells (1757–1817). See Warren, 'Darwin's missing links'.

# 7

## *Conclusion*

When the context of the natural history circles around and within the University of Edinburgh in the early decades of the nineteenth century are understood, it should not be surprising that it was here, rather than in anywhere else in Great Britain, that we have the richest evidence for the acceptance of the mutability of species. The secular, tolerant climate that had existed within the University provided a striking contrast to the paternalistic, religiously conservative regimes of the contemporary English universities. Both students and professors at Edinburgh came from much more varied social backgrounds than was true in the case of their English counterparts and were untroubled by any requirement to demonstrate their religious orthodoxy or loyalty to the Established Church. To a considerable extent they owed this openness to the tolerant academic regime established at the end of the previous century by the University's great principal William Robertson and his allies of the 'moderate literati' within and outside the University. This created a climate in which both professors and students could embrace and openly discuss new and exciting ideas about the natural world. While natural history at the University of Cambridge when Darwin was a student there from 1827 to 1831 was still viewed largely through the lens of William Paley's *Natural Theology* (1802), students at Edinburgh were encouraged to be more daring in their thinking on the natural world.[1] The records of the student societies provide ample evidence for lively debates on such controversial topics as materialistic theories of the human mind, the nature of organic life and, of course, the transmutation of species.

This atmosphere of free thought and open debate provided fertile ground for bold new theories that had their origin in continental Europe, and in France in particular. As a consequence of this openness, Edinburgh was integrated into a pan-European network for the exchange of ideas in a way that the English universities, still dominated by Paleyite natural theology, were not. It was commonplace for medical students and recent graduates from Edinburgh to spend time abroad at foreign universities where they picked up exciting new ideas. This was true of many of the key figures teaching natural history and related disciplines in Edinburgh in the 1820s. Both Robert Grant and Robert Knox had studied in Paris, the epicentre of debates on the transmutation of species in this period. Edinburgh's professor of natural history, Robert Jameson, had studied mineralogy with Werner in Freiberg. As a consequence of these connections, traces of the controversial theories of Lamarck and Geoffroy on the transmutation of species are to be found everywhere in the learned journals published in Edinburgh, books published by the city's natural historians, the minutes of its student societies and lectures of the University's professor of natural history, alongside echoes of the ideas of earlier thinkers, such as Erasmus Darwin, Buffon and Linnaeus.

As we have seen, the model of earth history embraced by the Wernerian geologists provided a convenient underpinning for a progressive, directional picture of the history of life, easily compatible with a model of the transmutation of species driven by constantly changing conditions. The Wernerian theory of the earth was radically different from the steady-state model of earth history espoused by the disciples of James Hutton. The Huttonian model assumed that the earth was in a state of dynamic equilibrium, while the Wernerians expected to find evidence of continuous, directional change in the geological record. Conditions on the surface of the earth in the past had been radically different from those of the present and would be different again in the future. It was an easy step from there to assuming, as Geoffroy had in Paris, that these changes in the physical conditions had driven the changes in living things observed in the fossil record. It is important to recognise that here the mechanism of change is fundamentally different from that proposed by Lamarck, who did not himself believe that conditions on the ancient earth had been radically different from the present. In his *Hydrogéologie* he had proposed his own essentially uniformitarian theory of the earth. For Lamarck the principal motor for progressive change was inherent in living things themselves rather than being driven by change in physical conditions. It would perhaps therefore be fairer to call the Edinburgh transformists 'Edinburgh Geoffroyans' rather than 'Edinburgh Lamarckians'.

While, as Pietro Corsi has pointed out, transformist theories could clearly provoke 'anxiety in moderate and conservative intellectual and scientific circles' across Europe before 1844, it is not clear to what extent it is possible to map the social and political tensions and conflicts of the period onto these scientific debates.[2] It sometimes seems too easy to assign transformism as the province of political and scientific radicals on the fringes of respectable society. However, the evidence presented in this study of the reception of transformism in Edinburgh strongly suggests that this neat equation is by no means reflected in the picture that emerges from a detailed analysis of contemporary sources. There is little solid evidence that many people prior to the publication of *Vestiges* were drawing political or social morals from the transmutation of species or claiming that transformism was a threat to the political and social order. The only concerted ideological opposition seems to have come from the evangelicals, who saw transformism as incompatible with some of the most important doctrines of their faith. Even the evangelicals were relatively restrained in their criticisms of Lamarck in the 1820s, and appear to have treated his theories with respect, although they generally ended up rejecting them as unduly speculative.

Some historians have painted a picture of a situation where transformism and transcendental anatomy in Britain were strongly associated with a radical faction in medical circles, among whom Grant played a central role. These figures are portrayed as in conflict with a medical establishment fiercely protective of its privileges.[3] However, the evidence presented in this study suggests that the Edinburgh context was quite different. In place of the hostile political camps of the London medical scene depicted by Adrian Desmond, what emerges from the surviving evidence from Edinburgh in the early decades of the nineteenth century is a surprisingly inclusive network of personal friendships and patronage extending across the extramural medical schools, the university and wider natural history circles in the city. When conflict occurred, which, as we have seen, it sometimes did, it was generally between individuals and for personal reasons, and cannot easily be interpreted as ideological conflict between representatives of entrenched political camps or interest groups. Despite the political and religious turmoil of the decades that followed the French Revolution, the Edinburgh natural history community remained surprisingly tolerant, open and cohesive in the 1820s.

Robert Grant is the figure for whom we have the strongest evidence in print from the mid-1820s onwards for the open advocacy of transformism. If the transmutation of species was a creed only espoused by radicals on the margins of the medical world at this time, we might expect Grant to have been a marginalised figure in Edinburgh and to have been treated with hostility by the professors of the University. This was clearly not the case. He appears to have been almost universally respected and admired, even by those natural historians who certainly did not share his belief in the transmutation of species. Overwhelming evidence for this is provided by the veritable constellation of luminaries from medical and natural history circles in Edinburgh, including David Brewster, John Fleming, Alexander Munro, tertius and Robert Jameson, who supported Grant's successful application for the post of professor of comparative anatomy at University College London in 1827. Jameson had also used his influence to advance the career of his former student by recommending him for membership of the Linnean Society of London in 1820. Fleming and Brewster were both leading members of the Evangelical Party of the Church of Scotland, and no friends to transformism, as was to be made abundantly clear in their reactions to the publication of *Vestiges of the Natural History of Creation* in 1844. However, they both seem to have maintained perfectly cordial relations with Grant in the 1820s. Brewster not only supported Grant's application for the chair at University College London, but also published a series of papers by Grant in the *Edinburgh Journal of Science*, including one of a decidedly transformist tendency in 1828. Grant seems to have been a particular friend of Fleming, who even did him the honour of naming a newly discovered species of sponge *Grantia*.

In our post-Darwinian world '[n]othing in biology makes sense except in the light of evolution', as Theodosius Dobzahansky famously entitled a 1973 essay.[4] It can sometimes be hard to remember that the same was not true in the early nineteenth century. While a controversial subject, the transmutation of species was one of many interlinked topics that stirred lively debate in the period. These also included materialistic theories of the human mind, the question of the vital principle and the debate over the rival theories of generation. Many natural historians in Edinburgh were clearly interested in theories of the transmutation of species, but it did not occupy centre stage in their debates as it was to do after 1859, or even after the publication of Chambers' *Vestiges* in 1844. Nobody in Edinburgh took the trouble to write a book devoted to the arguments either for or against the mutability of species, and such evidence as we have comes largely from journal articles, the records of society meetings and lecture notes. While theories of the transmutation of species seem to have been widely accepted as plausible, they appear to have been seen primarily as interesting speculations that did not conflict with a generally accepted image of the natural world as being in a state of continual progressive change and development, in accordance with the widely accepted Wernerian theory of the earth. Perhaps Grant and Cheek alone made the transmutation of species central to their visions of the natural world, although we have more limited evidence for the latter's views due to his early death. Transformist speculations were of interest to only a relatively small number of dedicated natural historians and only became genuinely controversial among a wider audience after the publication of *Vestiges* and the hostile reaction it elicited, especially from scientific Evangelicals such as David Brewster and Hugh Miller.

The publication of Robert Chambers' *Vestiges of the Natural History of Creation* in 1844 both revived and radically polarised the debate on the transmutation of species in the second half of the 1840s. The evolutionary theory elaborated by Chambers bears an indirect relation to those of the earlier Edinburgh transformists. The elements from which he built his theory were largely scavenged from the writings of physiologists who had studied in Edinburgh in the earlier decades of the nineteenth century. The books of John Fletcher and William Benjamin Carpenter in particular were mined by Chambers for the construction of his theory. His knowledge of the theories of Lamarck and Geoffroy seems to have been entirely second hand and derived from secondary sources, including Charles Lyell's hostile account in his *Principles of Geology*. Nor is it likely that Chambers was influenced directly by the ideas of the earlier Edinburgh transformists of the 1820s. Nonetheless, the building blocks from which he developed his theory, and above all his extensive reliance on the work of John Fletcher, speak of the important debt his theory owed to the intellectual climate of Edinburgh's natural history circles in the 1820s and in particular the strong influence of Geoffroy's theories.

While never a secret creed, from the point of view of mainstream natural history in Britain in the early nineteenth century theories of the transmutation of species formed a largely subterranean stream, the momentous implications of which would become apparent only much later. While these ideas were much more widely and openly debated on the broader European stage, in England at least the firm grip of natural-theological thinking and a strong prejudice against continental ideas in the aftermath of the French Revolution largely held them at bay. Edinburgh, with its strong continental connections, seems to have been a striking exception to this picture. The natural history circle around Robert Jameson was the main conduit through which this stream of new ideas flowed into British natural history. From there they were spread more widely, often in a modified or developed form, by figures such as Grant, Knox, Watson and the generations of medical students who had studied in Edinburgh and attended Jameson's classes in the 1820s. As we have seen in the previous chapter, when Charles Darwin, who had been one of those students in the late 1820s, turned his mind seriously to the transmutation of species after his return from the voyage of the *Beagle*, he was still largely following the Grantian programme, informed by the theories of Lamarck and Geoffroy. Elements of that programme were to form the bedrock on which he was to build his own theory of evolution. Darwin, in his determination to present his theory as a model piece of inductive science, must take some of the responsibility for obscuring the roots of his ideas. The theories of transformist thinkers such as Lamarck were considered scandalously speculative and methodologically unsound in mid-century England, where Bacon was held up as the patron saint of correct scientific method. Acknowledging a debt to them and their Edinburgh disciples would not have furthered his ambitions for himself or his theory, but would rather have been likely to do lasting damage to his reputation. So in his published work he generally passed over in silence the debt that he had acknowledged more openly in his private notebooks.

Much time and energy has been devoted by historians of science over the years in tracking down the precursors of Darwin and assessing their influence on him, and a vast amount of light has thus been shed on the roots of his theories. In this study of transformism in early nineteenth-century Edinburgh I have tried to avoid taking this approach, partly because so many eminent scholars have travelled that path before, but also in the belief that when studying the transformists of this period it may be more profitable to try to understand them on their own terms and in their own context, setting aside any question of their later influence. If we can resist the temptation to see early nineteenth-century transformists primarily as precursors of Darwin, but rather view them as heirs to a tradition of progressive theorising on the history of the earth and of life that reaches far back into the eighteenth century and was irrigated from the wellsprings of Enlightenment optimism, we may ultimately gain a clearer understanding of their true significance. We are beginning to see to what extent transformist ideas were not only the province of shadowy figures on the fringes of science and society, but formed one of the competing strands of mainstream thought on the broader European stage.

## NOTES

1. Although, as Aileen Fyfe has established, Paley's *Natural Theology* was not in fact a set text at this time. See Fyfe, 'Reception of William Paley's *Natural Theology*'.
2. Corsi, 'Lamarck: From myth to history', p. 16.
3. See in particular Desmond, *Politics of Evolution*.
4. Dobzhansky, 'Nothing in biology makes sense except in the light of evolution'.

# Bibliography

## UNPUBLISHED PRIMARY SOURCES

Anon, Fragments of a translation of Jean-Baptiste Lamarck, *Histoire naturelle des animaux sans vertèbres*, vols 1–6.1 (1815–22), Jameson papers (Edinburgh University Library, Gen.124).

Chambers, Robert to Combe, George, 14 December 1833, Chambers papers (National Library of Scotland, Ms.7330).

Chambers, Robert to Combe, George, 25 November 1835, Chambers papers (National Library of Scotland, Ms.7234).

Chambers, Robert to Ireland, Alexander [day and month unknown] 1847 (National Library of Scotland, Dep. 341/110).

Chambers, Robert to Ireland, Alexander, 29 March 1845 (National Library of Scotland, Dep. 341/113).

Cheek, Henry, 'On the varieties of the human race', The Royal Medical Society, Dissertations 91 (1829–30), 286–307 (Library of the Royal Medical Society of Edinburgh).

Combe, George to Robert Chambers, 26 November 1835, Combe papers (National Library of Scotland, Ms.7386).

Combe, George to Chambers, Robert, 13 December 1833, Chambers papers (National Library of Scotland, Ms.7386).

Combe, George to Chambers, Robert, 15 December 1835, Robert Chambers, letters of noted persons, 1833–38 (National Library of Scotland, Dep.341/91).

Gilby. H., 'Question on the comparative merits of the Huttonian and Wernerian theories of the earth', The Royal Medical Society, Dissertations 70 (1813/14), 439–55 (Library of the Royal Medical Society of Edinburgh).

Grant, Robert, Essays on medical subjects (date unknown) (UCL library, MS ADD 28 (box 17)).

Grant, Robert, Notes on 'Lectures on comparative anatomy delivered in the University of London by Robert E Grant M.D. F.R.S. &c. Session 1833 1834' (anonymous student) (UCL Library, MS ADD 38 (box 19)).

[Jameson, Robert], Untitled manuscript on the transmutation of species (date unknown) (Jameson papers, Edinburgh University Library, Gen.125).

Jameson, Robert, Notes on natural history lectures (1806) (taken by John Borwick) (Edinburgh University Library, Gen.847).

Jameson, Robert, Notes on natural history lectures (watermark 1813/14) (taken by William Dansey), 2 vols (Edinburgh University Library, Dc.3.33–4).

[Jameson, Robert], Note on the verso of a card bearing the inscription 'Civis Bibliotecae Academiae Edinburgenae a die 12 Octobris 1824, ad diem 12 Octobris 1825' (Edinburgh University Library, Gen. 1999/2/3).

Jameson, Robert, Notes on natural history lectures (1830) (taken by W. S. Walker) (National Library of Scotland, Ms.14148).

Jameson, Robert, Student notes of Jameson's lectures on natural history (1830/1) (taken by R. M'Cormick) (Wellcome Library, Ms.3358).

Jameson, Robert, Notes on natural history lectures (anonymous student) (1831/2) (National Library of Scotland, Ms.3936).

Jameson, Robert, Notes on natural history lectures, 3 vols (1835/6) (taken by David Blair Ramsay) (Glasgow University Library, MS Cullen 281–3).

McDowell, John C., 'An examination of the igneous or Huttonian theory of the earth', The Royal Physical Society, Dissertations 25 (1806/7), 313–29 (Edinburgh University Library, Da.67 Phys).

Monro, Alexander, secundus, Notes on lectures on anatomy (1774/5) (taken by James Johnson), 4 vols (Edinburgh University Library, Gen.569–73).

Monro, Alexander, secundus, Notes on lectures on comparative anatomy (1786) (anonymous student) (Edinburgh University Library, Dc.10.13)

Ogilvy, J. 1806/7, 'On the Huttonian and Neptunian theories of the earth', The Royal Medical Society, Dissertations [56] (1806/7), 230–48. Library of the Royal Medical Society of Edinburgh.

Plinian Natural History Society, Minutes of the Plinian Society, 2 vols (Edinburgh University Library, Dc.2.53–4).

Spurzheim, Johann to Combe, George, 31 December 1831 (National Library of Scotland, Ms.7207).

Treuttel & Co. receipts, Jameson papers (Edinburgh University Library, Gen.130).

Stewart, Ralph Smyth, 'On the different theories of the earth', Royal Physical Society, Dissertations [35] (1821–2), 493–508 (Edinburgh University Library, Da.67 Phys).

Walker, John, Notes on natural history lectures (1782) (taken by Charles Stewart) (Edinburgh University Library, Dc.2.22).

Walker, John, Notes on natural history lectures (1790) (anonymous student), 6 vols (Edinburgh University Library, Dc.2.25–8).

Walker, John, Notes on natural history lectures (1791) (anonymous student) (Edinburgh University Library, Dc.10.33).

Walker, John, Notes on natural history lectures (1797) (taken by David Pollock), 10 vols (Edinburgh University Library, Gen.703).

Watson, Hewett Cottrell to Combe, George, 14 December 1836 (Edinburgh University Library Ms.7241).

Watson, J. William, 'Essay on petrifactions', The Royal Physical Society, Dissertations 27 (1806–13), 202–26 (Edinburgh University Library, Da.67 Phys).

Wernerian Natural History Society, Minutes of the Wernerian Society, 1808–58, 2 vols (Edinburgh University Library, Dc.2.55–6).

## PUBLISHED PRIMARY SOURCES

Ainsworth, William Francis, 'A descent into Eldon Hole, in the Peak of Derbyshire', *Ainsworth's Magazine: A Miscellany of Romance, Literature and Art* 2 (1842), 259–63.
Anon, 'Some account of the Wernerian Natural History Society of Edinburgh', *Blackwood's Edinburgh Magazine* 3: 1 (1817), 231–4.
Anon, 'Observations on the nature and importance of geology', *Edinburgh New Philosophical Journal* 1 (1826), 293–302.
Anon, 'Of the changes which life has experienced on the globe', *Edinburgh New Philosophical Journal* 3 (1827), 298–301.
Anon, *Abstract of the Proceedings of the Plinian Society from its first meeting Jan 14, 1823, to July 25, 1826* (Edinburgh: MacLachlan, Stewart & Co, 1829).
Anon, 'Dinner by the Phrenological Society to Dr Spurzheim', *Phrenological Journal* 5 (1829), 102–42.
Anon, 'Natural history in Scotland', *The Magazine of Natural History and Journal of Zoology, Botany, Mineralogy, Geology, and Meteorology*, 1 (1829), 291–2.
Anon, 'Of the continuity of the animal kingdom by means of generation from the first ages of the world to the present times: On the relations of organic structure and parentage that may exist between animals of the historic ages and those at present living, and the antediluvian and extinct species', *Edinburgh New Philosophical Journal* 7 (1829), 152–6.
Anon, 'Science: Académie des Sciences. Séance du lundi 23 mars 1829', *Le Globe* 7: 26 (1829), pp.207–8.
Anon, 'Remarks on the ancient flora of the Earth', *Edinburgh New Philosophical Journal* 8 (1830), 112–31.
Anon, 'Query on the hereditary transmission of accidental characters', *Edinburgh Journal of Natural and Geographical Science* 3 (March 1831), 173.
Anon, 'Local scientific societies', *Nature* 9 (1873), 38–40.
Babbage, Charles, *Reflections on the Decline of Science in England and Some of its Causes* (London: B. Fellowes, 1830).
Baird, William, *Memoir of the Late Rev. John Baird, Minister of Yetholm, Roxburghshire; with an Account of his Labours in Reforming the Gypsy Population of that Parish* (London: James Nisbet & Co., 1862).
Balfour, John Hutton, *Biographical Sketch of the Late John Coldstream* (Edinburgh: T. Constable, 1864).
Barclay, John, *New Anatomical Nomenclature, Relating to the Terms which are Expressive of Position and Aspect In the Animal System* (Edinburgh: Ross and Blackwood, 1803).
Barclay, John, *An Enquiry in to the Opinions, Ancient and Modern, Concerning Life and Organization* (Edinburgh: Bell & Bradfute, 1822).

Barclay, John, *Introductory Lectures to a Course of Anatomy, delivered by John Barclay, M.D. F.R.S.E. with a Memoir of the Life of the Author* (Edinburgh: MacLachlan and Stewart, 1827).

Charles Bonnet, *Contemplation de la Nature* (Amsterdam: Marc-Michel Rey, 1764).

Boué, Ami, 'Geological observation, – 1. On alluvial rocks: 2. On formations: 3. On the changes that appear to have taken place during the different periods of the Earth's formation on the climate of our Globe, and in the nature and the physical and the geographical distribution of its animals and plants', *Edinburgh New Philosophical Journal* 1 (1826), 82–92.

Boué, Ami, *Bulletin de la Société Géologique de France: Résumé des progrès de sciences géologiques pendant l'année 1833* (Paris: Société Géologique de France, 1834).

Boué, Ami, 'Autobiographie pour mes amis', in *Autobiographie du Docteur Medecin Ami Boue: Et catalogue des œuvres, travaux, memoires et notices* (Vienna: Ulrich, 1879).

Bower, Alexander, *The Edinburgh Student's Guide: Or an Account of the Classes of the University, Arranged under Four Faculties; with a Detail of What Is Taught in Each* (Edinburgh: Waugh and Innes, 1822).

[Brewster, David], 'Advertisement', *Edinburgh Journal of Science* 1 (1824), viii–ix.

[Brewster, David], Review of Charles Babbage's *Reflections on the Decline of Science in England, and Some of Its Causes*, *The Quarterly Review* 43: 86 (1830), 305–42.

[Brewster, David], Review of [Robert Chambers'] *Vestiges of the Natural History of Creation*, *North British Review*, 3: 6 (1845), 470–515.

Brougham, Henry Lord, *Life and Times of Henry Lord Brougham Written by Himself*, 3 vols (Edinburgh: William Blackwood and Sons, 1871).

Buckland, William, *Geology and Mineralogy Considered with Reference to Natural Theology*, 2 vols (London: William Pickering, 1836).

Buffon, Georges-Louis Leclerc, Comte de, *Histoire Naturelle, Générale et Particulière, supplément*, 7 vols (Paris: L'Imprimerie Royale, 1774–89).

Buffon Georges-Louis Leclerc, Comte de, *Natural History, General and Particular*, 2nd edn, 9 vols (trans William Smellie) (London: W. Straham and T. Cadell, 1785).

Carpenter, William Benjamin, *Principles of General and Comparative Physiology*, 2nd edn (London: John Churchill, 1841).

[Chambers, Robert], 'Natural history: Animals with a backbone', *Chambers's Edinburgh Journal* 1: 43 (Saturday, 24 November 1832), 337–8.

[Chambers, Robert], 'Natural history: Monkeys, apes, and orang-outangs' *Chambers's Edinburgh Journal* 1: 46 (Saturday, 15 December 1832), 362.

[Chambers, Robert], 'Is ignorance bliss?', *Chambers's Edinburgh Journal* 2: 49 (4 January 1834), 385–6.

[Chambers, Robert], 'Popular information on science: Transmutation of species',

*Chambers's Edinburgh Journal*, vol.4, No.191 (Saturday, 26 September 1835), 273–4.

[Chambers, Robert], 'Popular information on science: Third ages of animal life', *Chambers's Edinburgh Journal* 6: 298 (Saturday, 14 October 1837), 298–9.

[Chambers, Robert], 'The life and poetry of Darwin', *Chambers's Edinburgh Journal*, 8: 394 (Saturday, 17 August 1839), 251.

[Chambers, Robert], *Vestiges of the Natural History of Creation* (London: John Churchill, 1844).

Chambers, Robert, *Ancient Sea-margins, as Memorials of Changes in the Relative Level of Sea and Land* (Edinburgh: W. & R. Chambers, 1848).

[Cheek, Henry], 'On the natural history of the Dugong (*Halicore Indicus*, Desm.) – the mermaid of early writers, and particularly on the differences which occur in its dental characters', *Edinburgh Journal of Natural and Geographical Science* 1 (1829), 161–72.

[Cheek, Henry], 'Review of the recent discussion before the Academy of Sciences in Paris, on the "unity of organization". Part I,– Baron Cuvier's views', *Edinburgh Journal of Natural and Geographical Science* 2 (1830), 37–40.

[Cheek, Henry], 'Suggestions on the relation between organized bodies, and the conditions of their existence', *Edinburgh Journal of Natural and Geographical Science* 2 (1830), 65.

[Cheek, Henry], 'On the present state of science abroad: No.1 Scientific coteries of Paris', *Edinburgh Journal of Natural and Geographical Science* 2 (1830), 116–20.

Cheek, Henry, *An Answer to Certain Statements Contained in Mr Neill's 'Address to the Members of the Wernerian Natural History Society. 25 September 1830* (Edinburgh, 1830).

[Cheek, Henry], 'On the existence of vascular arches in the foetus of mammifera, birds, and reptiles, similar to the branchial arteries in fishes and the larvae of the batrachian reptiles', *Edinburgh Journal of Natural and Geographical Science* 3 (1831), 235–8.

[Cheek, Henry], 'Miscellaneous intelligence: Edinburgh University', *Edinburgh Journal of Natural and Geographical Science* 3 (1831), 77.

[Cheek, Henry], Editorial comment on 'Query on the hereditary transmission of accidental characters', *Edinburgh Journal of Natural and Geographical Science* 3 (March 1831), 173.

Cockburn, Henry, *Memorials of His Time* (Edinburgh: Adam and Charles Black, 1856).

Cockburn, Henry to Kennedy, Thomas Francis, 27 February 1826, in Cockburn, Henry and Kennedy, Thomas Francis, *Letters Chiefly Concerned with the Affairs of Scotland from Henry Cockburn to Thomas Francis Kennedy* (London: William Ridgway, 1874).

Combe, George, *Elements of Phrenology*, 3rd edn (Edinburgh: John Anderson, 1828).

Combe, George, *A System of Phrenology*, 3rd edn (Edinburgh: John Anderson, 1830).
Combe, George, *The Constitution of Man Considered in Relation to External Objects*, 4th edn (Edinburgh: William and Robert Chambers, 1835).
Cunningham, Robert James Hay, 'On the geology of the Lothians', *Memoirs of the Wernerian Natural History Society* 7 (1831–37), 3–160.
Cuvier, Georges, *Recherches sur les ossements fossiles de quadrupèdes*, 4 vols (Paris: Deterville, 1812).
Cuvier, Georges, *Essay on the Theory of the Earth*, 1st edn (trans. Robert Kerr) (Edinburgh: William Blackwood, 1815).
Cuvier, Georges, *Le Règne Animal Distribué d'après son Organisation*, 4 vols (Paris: Deterville, 1817).
Cuvier, Georges, *Discours sur les Révolutions de la surface du globe*, 3rd edn (Paris: G. Defour et Ed. D'Ocagne, 1825).
Cuvier, Georges, *Essay on the Theory of the Earth*, 5th edn (trans. Robert Kerr and Robert Jameson) (Edinburgh: William Blackwood, 1827).
Cuvier, Georges, 'Memoir of M. de Lamarck', *The Edinburgh New Philosophical Journal* 20: 39 (1836), 1–22.
Darwin, Charles to Darwin, Caroline, 6 January 1826, in Darwin, Charles, *The Correspondence of Charles Darwin, Vol. 1: 1821–1836* (Cambridge: Cambridge University Press, 1985).
Darwin, Charles to Henslow, John Stevens, 18 May 1832, *The Correspondence of Charles Darwin, Vol. 1: 1821–1836* (Cambridge: Cambridge University Press, 1985).
Darwin, Charles, Notebook B [Transmutation of species (1837–1838)]. (Cambridge University Library, DAR121) (transcribed by Kees Rookmaaker), in Van Wyhe, John (ed.) 2002–. *The Complete Work of Charles Darwin Online*. <http://darwin-online.org.uk/> (last accessed 6 August 2016).
Darwin, Charles, Notebook C [Transmutation of species (1838.02–1838.07)] (Cambridge University Library, DAR121) (transcribed by Kees Rookmaaker), in Van Wyhe, John (ed.) 2002–. *The Complete Work of Charles Darwin Online*. <http://darwin-online.org.uk/> (last accessed 6 August 2016).
Darwin, Charles, Notebook D [Transmutation of species (7–10.1838)] (Cambridge University Library, DAR123) (transcribed by Kees Rookmaaker), in Van Wyhe, John (ed.) 2002–. *The Complete Work of Charles Darwin Online*. <http://darwin-online.org.uk/> (last accessed 6 August 2016).
Darwin, Charles, 'The essay of 1844', in Darwin, Charles, *The Foundations of The Origin of Species. Two Essays written in 1842 and 1844* (ed. Francis Darwin) (Cambridge: Cambridge University Press, 1909).
Darwin, Charles to Hooker, J. D., [29 May 1854], Darwin Correspondence Database, <www.darwinproject.ac.uk/entry-1575> (last accessed on 19 March 2013).
Darwin, Charles, *Autobiographies* (ed. Michael Neve) (London: Penguin, 2002).

De Candolle, Alphonse, 'On the history of fossil vegetables', *Edinburgh New Philosophical Journal* 18 (1835), 81–102.
Duncan, James, 'Memoir of Lamarck', in Jardine, William, *The Naturalist's Library. Entomology. Vol. V. Foreign Butterflies* (Edinburgh: W. H. Lizars, 1837).
Esmark, Jens, 'Remarks tending to explain the geological history of the Earth', *Edinburgh New Philosophical Journal* 2 (1827), 107–21.
[Fleming, John], Review of Jean-Baptiste Lamark, *Histoire naturelle des animaux sans vertèbres*, *Edinburgh Review* 3: 4 (April 1820), 403–18.
Fleming, John, *The Philosophy of Zoology or a General View of the Structure, Functions, and Classification of Animals*, 2 vols (Edinburgh: Archibald Constable & Co., 1822).
Fleming, John, *A History of British Animals, Exhibiting the Descriptive Characters and Systematical Arrangement of the Genera and Species of Quadrupeds, Birds, Reptiles, Fishes, Mollusca, and Radiata of the United Kingdom* (Edinburgh: Bell & Bradfute, 1828).
[Fleming, John], Review of J. E. Bicheno, 'On systems and methods in natural history', *The Quarterly Review* 42 (November 1829), 302–27.
Fleming, John, *Inauguration of the New College of the Free Church, Edinburgh, November, M.DCCC.L. with Introductory Lectures on Theology, Philosophy and Natural Science* (Edinburgh: Johnstone and Hunter, 1851).
Fleming, John, *The Lithology of Edinburgh* (Edinburgh: William P. Kennedy, 1859).
Fletcher, John, *Rudiments of Physiology, in Three Parts* (Edinburgh: William P. Kennedy, 1835).
Forbes, Edward, 'Professor Forbes's inaugural lecture', *The Scotsman*, 17 May 1854, 4.
Geoffroy Saint-Hilaire, Étienne, 'Recherches sur l'organisation des gavials; Sur les affinités naturelles desquelles résulte la nécessité d'une autre distribution générique, Gavialis, Teleosaurus et Steneosaurus; et sur cette question, si les Gavials (Gavialis), aujourd'hui répandus dans les parties orientales de l'Asie, descendent, par voie non interrompue de génération, des Gavials antidiluviens, soit des Gavials fossiles, dits Crocodiles de Caen (Teleosaurus), soit des Gavials fossiles du Havre et de Honfleur (Stenosaurus)', *Mémoires du Muséum d'Histoire Naturelle* 12 (1825), 97–155.
Geoffroy Saint-Hilaire, Étienne, 'Anencéphales humains', *Mémoires du Muséum d'Histoire Naturelle* 12 (1825), 257–92.
Geoffroy Saint-Hilaire, Étienne and Serres, Étienne, 'Rapport fait à l'Académie Royale des Sciences sur une mémoire de M. Roulin, ayant pour titre : *Sur quelques changemens observés dans les animaux domestiques transportés de m'ancien monde dans le nouveau continent*', *Mémoires du Muséum d'Histoire Naturelle* 17 (1828), 201–8.
Geoffroy Saint-Hilaire, Étienne, 'Mémoire: Où l'on propose de rechercher dans quelles rapports de structure organique et de parenté sont entre eux les animaux

des âges historiques, et vivant actuellement, et les espèces antédiluviennes et perdues', *Mémoires du Muséum d'Histoire Naturelle* 26 (1828), 209–29.

Geoffroy Saint-Hilaire, Étienne, *Recherches sur de grands sauriens trouvés a l état fossile vers les confins maritimes de la Basse Normandie, attribués d'abord au crocodile, puis déterminés sous les noms de téléosaurus et sténéosaurus* (Paris: Firmin Didot Frères, 1831).

Geoffroy Saint-Hilaire, Étienne, 'Quatrième mémoire, lu à l'académie des sciences, le 28 mars 1831', in Geoffroy Saint-Hilaire, Étienne, *Recherches sur de grands sauriens* (Paris: Firmin Didot Frères, 1831).

Geoffroy Saint-Hilaire, Étienne, 'Palaeontographie : Considérations sur des ossemens fossiles la plupart inconnus, trouvés et observés dans les bassins de l'Auvergne', in Carnot, H. and Leroux, P. (eds), *Revue Encyclopédique*, vol.59 (Paris: Bureau de la Revue Encyclopédique, 1833).

Geoffroy Saint-Hilaire, Étienne, *Études progressives d'un naturaliste pendant les années 1834 et 1835* (Paris: Roret, 1835).

Geoffroy Saint-Hilaire, Étienne, 'Preliminary discourse', from *Anatomical Philosophy: On the Respiratory Organs with Respect to the Determination and the Identity of their Bony Parts* (1818), in le Guyader, Hervé, *Geoffroy Saint-Hilaire: A Visionary Naturalist* (trans. Marjorie Grene) (Chicago, IL: University of Chicago Press, 2004), 26–35.

Geoffroy Saint-Hilaire, Étienne, 'First memoir', from *Memoirs on the Organization of Insects* (1820), in le Guyader, Hervé, *Geoffroy Saint-Hilaire: A Visionary Naturalist* (trans. Marjorie Grene) (Chicago, IL: University of Chicago Press, 2004), 53–63.

Étienne Geoffroy Saint-Hilaire, 'General considerations on the vertebra' (1822), in le Guyader, *Geoffroy Saint-Hilaire: A Visionary Naturalist* (trans. Marjorie Grene) (Chicago, IL: University of Chicago Press, 2004), 64–87.

Gordon, Margaret Maria, *The Home Life of Sir David Brewster* (Edinburgh: Edmonston and Douglas, 1869).

Grant, Alexander, *The Story of the University of Edinburgh during its First Three Hundred Years*, 2 vols (London: Longmans, Green and Co., 1884).

Grant, Robert, *Dissertatio Physiologica Inauguralis, de Circuitu Sanguinis in Foetu* (Edinburgh : Jac. Ballantyne et Socii, 1814).

Grant, Robert, 'On the structure and nature of the Spongilla friabilis', *Edinburgh Philosophical Journal* 14 (1826), 270–84.

Grant, Robert, 'Observations on the structure of some silicious sponges', *Edinburgh New Philosophical Journal* 1 (1826), 341–51.

Grant, Robert, 'Notice regarding the ova of the Pontobdella muricata', *Edinburgh Journal of Science* 7 (1827), 160–2.

Grant, Robert, 'Observations on the generation of the Lobularia digitata, Lam. (Alcyonium lobatum, Pall.)', *Edinburgh Journal of Science* 8 (1828), 104–10.

Grant, Robert, *An Essay on the Study of the Animal Kingdom: Being an Introductory Lecture Delivered in the University of London, on the 23rd of October*, 1828 (London: John Taylor, 1829).

Grant, Robert, *Comparative Anatomy and Zoology* (syllabus) (London, 1830).
Grant, Robert, 'University of London lectures on comparative anatomy and animal physiology: Lecture VI. On the organs of support of acephala and echinoderma', *The Lancet* 1 (1833–4), 265–79.
Grant, Robert, *Tabular View of the Primary Divisions of the Animal Kingdom, Intended to Serve as an Outline of an Elementary Course of Recent Zoology* (London: Walton and Maberly, 1861).
Grierson, James, 'General observations on geology and geognosy, and the nature of these respective studies', *Memoirs of the Wernerian Natural History Society* 5 (1823–24), 401–10.
Hall, James, *Account of a Series of Experiments, Shewing the Effects of Compression in Modifying the Action of Heat* (Edinburgh, 1805).
Hibbert Ware, Mary Clementina, *The Life and Correspondence of the Late Samuel Hibbert Ware* (Manchester: J. E. Cornish, 1882).
Hume, A., *The Learned Societies and Printing Clubs of Great Britain* (London: G. Willis, 1853).
Hutton, James, 'Theory of the earth; or an investigation of the laws observable in the composition, dissolution and restoration of land upon the globe', *Transactions of the Royal Society of Edinburgh* 1 (1788), 209–304.
Jameson, Laurence, *Biographical Memoir of the Late Professor Jameson* (Edinburgh: Neill and Company, 1854).
Jameson, Robert, *System of Mineralogy*, 3 vols (Edinburgh: Archibald Constable and Co., 1804).
Jameson, Robert, *Elements of Geognosy* (Edinburgh: William Blackwood, 1808).
[Jameson, Robert], 'New publications received', *Edinburgh New Philosophical Journal* 38 (1845), 186–8.
[Jameson, Robert], 'New publications received', *Edinburgh New Philosophical Journal* 40 (1846), 399–401.
Jardine, George, *Outlines of Philosophical Education, Illustrated by the Method of Teaching the First Class of Philosophy in the University of Glasgow* (Glasgow: Oliver & Boyd, 1825).
Johnston, George, 'A few remarks on the class Mollusca, in Dr Fleming's work on British animals; with descriptions of some new species', *Edinburgh New Philosophical Journal* 5 (1828), 74–81.
Herschel, John Frederick William, *A Preliminary Discourse on the Study of Natural Philosophy* (London: Longman, Rees, Orme, Brown and Green, 1830).
Kay, John, *A Series of Original Portraits and Caricature Etchings, by the Late John Kay, Miniature Painter, Edinburgh; With Biographical Sketches and Anecdotes*, 2 vols (Edinburgh: Hugh Paton, 1838).
Knox, Robert, 'An account of the *Foramen centrale* of the retina generally called the *Foramen of Soemmering*, as seen in the eyes of certain reptiles', *Memoirs of the Wernerian Society* 5: 1 (1824), 1–7.
Knox, Robert, 'Observations on the duck-billed animal of New South Wales, the

*Ornithorynchus paradoxus* of naturalists: Memoir I. On the organs of sense, and on the anatomy of the poison gland and spur', *Memoirs of the Wernerian Society* 5: 1 (1824), 26–41.
Knox, Robert, 'Additional observations relative to the *foramen centrale* of the retina in reptiles', *Memoirs of the Wernerian Society* 5: 1 (1824), 104–6.
Knox, Robert, 'Inquiry into the origin and characteristic differences of the native races inhabiting the extra-tropical part of southern Africa', *Memoirs of the Wernerian Society* 5: 1 (1824), 206–19.
Knox, Robert, 'Notice respecting the presence of a rudimentary spur in the female echidna of New Holland', *Edinburgh New Philosophical Journal* 1 (1826), 130–2.
Knox, Robert, 'Lectures of M. De Blainville on comparative osteology. The comparative osteography of the skeleton and dentary system, in the five classes of vertebral animals, recent and fossil, by M.H.M. Ducrotay de Blainville', *The Lancet* 1 (26 October 1839), 137–45.
Knox, Robert, *Great Artists and Great Anatomists: A Biographical and Philosophical Study* (London: John van Voorst, 1852).
Knox, Robert, 'Enquiries into the philosophy of zoology. Part I. – On the dentition of Salmonidae', *The Zoologist: A Popular Miscellany of Natural History* 13 (1855), 4777–92.
Knox, Robert, 'Introduction to Enquiries into the philosophy of zoology', *The Lancet* 7 (23 June 1855), 625–7.
Knox, Robert, 'The philosophy of zoology, with special reference to the natural history of man', *The Lancet* 7 (14 July 1855), 216–18
Knox, Robert, *The Races of Men: A Fragment* (London: Henry Renshaw, 1850).
Lamarck, Jean-Baptiste, 'Prodrome d'une nouvelle classification des Coquilles', *Mémoires de la Société d'Histoire Naturelle de Paris* 1 (1799), 67–91.
Lamarck, Jean-Baptiste, *Hydrogéologie* (Paris: Agasse, 1802).
Lamarck, Jean-Baptiste, 'Sur les fossiles des environs de Paris, comprenant des espèces qui appartiennent aux animaux marins sans vertèbres, dont la plupart sont figurés dans la collection des vélins du Muséum', *Annales du Muséum d'Histoire Naturelle* 1 (1802), 299–307.
Lamarck, Jean-Baptiste, *Recherches sur l'organisation des corps vivants* (Paris: Maillard, 1802).
Lamarck, Jean-Baptiste, *Philosophie zoologique*, 2 vols (Paris: Dentu, 1809).
Lamarck, Jean-Baptiste, *Histoire naturelle des animaux sans vertèbres*, 7 vols (Paris: Verdière, 1815–22).
Leslie, John, *An Experimental Inquiry into the Nature and Propagation of Heat* (London: J. Mawman, 1804).
[Lockhart, J. G.], *Peter's Letters to his Kinsfolk*, 3 vols (Edinburgh: William Blackwood, 1819).
Lonsdale, Henry, *A Sketch of the Life and Writings of Robert Knox the Anatomist* (London: Macmillan and Co., 1870).
Lord, Percival B., *Popular Physiology; Being a Familiar Explanation of the Most*

*Interesting Facts Connected with the Structure and Function of Animals and Particularly Man* (London: John W. Parker, 1839).
Lorimer, James, *The Universities of Scotland Past, Present and Possible* (Edinburgh: W.P. Kennedy, 1854).
Lyell, Charles, *Principles of Geology*, 3 vols (London: John Murray, 1830–33).
Mackintosh, James, *Memoirs of the Life of the Right Honourable Sir James Macintosh* (edited by Robert James Mackintosh) (London: Edward Moxon, 1836).
Maupertuis, Pierre-Louis, *Les Œuvres de Monsieur de Maupertuis* (Dresden: George Conrad Walther, 1752).
Miller, Hugh, *The Old Red Sandstone; or New Walks in an Old Field* (Edinburgh: John Johnstone, 1841).
Miller, Hugh, *Foot-Prints of the Creator: or the Asterolepis of Stromness* (London: Johnstone and Hunter, 1849).
[Mudie, R.], *The Modern Athens: A Dissection and Demonstration of Men and Things in the Scotch Capital* (London: Knight and Lacey, 1824).
[Muirhead, Lockhart], Review of Lamarck's *Philosophie zoologique*, Monthly Review 55 (August 1811), 473–84.
Neill, Patrick, *Supplement to an Address to the Wernerian Natural History Society, Dated July 1830; Containing a Reply by Mr Neill to Mr Cheek's Answer, November 1830* (Edinburgh, 1830).
Nichol, John Pringle, *Views of the Architecture of the Heavens in a Series of Letters to a Lady* (Edinburgh: Neill and Company, 1837).
Playfair, John, *Illustrations of the Huttonian Theory of the Earth* (Edinburgh: Cadell and Davies, 1802).
Plinian Natural History Society, *Abstract of the Proceedings of the Plinian Society from its First Meeting Jan 14, 1823, to July 25, 1826* (Edinburgh: Maclachlan, Stewart & Co., 1829).
Alexander Pope, *An Essay on Man* (Glasgow: Alexander Weir, 1768).
Powell, Baden, *Essays on the Spirit of the Inductive Philosophy, the Unity of Worlds, and the Philosophy of Creation* (London: Longman, Brown, Green and Longman, 1855).
Royal Medical Society, *List of Members, The Royal Medical Society of Edinburgh* (Edinburgh: Royal Medical Society, 1906).
Scott, William, *The Harmony of Phrenology with Scripture: Shewn in a Refutation of the Philosophical Errors Contained in Mr Combe's 'Constitution of Man'* (Edinburgh: Fraser & Co., 1836).
Scottish Universities Commission (1826), *General Report of the Commissioners Appointed to Visit the Universities and Colleges of Scotland* (London: HM Stationery Office, 1830).
Scottish Universities Commission (1826), *Minutes of Evidence Taken before the Commissioners for Visiting the Universities and Colleges in Scotland: University of Edinburgh 1826, 1827, 1830* (Edinburgh, 1830).

Scottish Universities Commission (1826), *Report Relative to the University of Edinburgh* (Edinburgh, 1830).
Scottish Universities Commission (1826), *Returns, Papers, and Examinations Printed by Order of the Royal Commissioners for Visiting the Universities and Colleges of Scotland* (Edinburgh, 1830?).
Scottish Universities Commission, *An Abstract of the General Report of the Royal Commissioners Appointed to Visit the Universities of Scotland with Notes and Tabular States Relating to the State of these Institutions in 1826* (Edinburgh: Adam and Charles Black, 1836).
Scottish Universities Commission (1826), *Evidence, Oral and Documentary, Taken and Received by the Commissioners Appointed by His Majesty George IV., July 23d, 1826; and Re-appointed by His Majesty William IV., October 12th, 1830; for Visiting the Universities of Scotland. Volume 1. University of Edinburgh* (London: HM Stationery Office, 1837).
Sedgwick, Adam, *A Discourse on the Studies of the University*, 2nd edn (Cambridge: J. & J. J. Deighton, 1834).
Shuttleworth, B. S., 'Hereditary transmission of accidental characters', *Edinburgh Journal of Natural and Geographical Science* 3 (1831), 301.
Smellie, William, *The Philosophy of Natural History* (Edinburgh: C. Elliot and T. Kay, 1790).
Smollett, Tobias, *The Expedition of Humphrey Clinker* (Oxford: Oxford University Press, 1998).
Stewart, Dugald, 'Account of the life and writings of William Robertson, DD', in Robertson, William, *The Works of William Robertson, DD*, vol.1 (London: T. Cadell, 1827).
Tiedemann, Frederick, *A Systematic Treatise on Comparative Physiology, Introductory to the Physiology of Man* (trans. James Manby Gully and James Hunter Lane) (London: John Churchill, 1834).
[Wakley, Thomas], 'Biographical sketch of Robert Edmond Grant, M.D. F.R.S.L. & E. &c. Professor of Comparative Anatomy and Zoology in University College, London', *The Lancet* 56 (21 December 1850), 686–95.
Walker, John, *Institutes of Natural History: Containing the Heads of the Lectures in Natural History, Delivered by Dr Walker, in the University of Edinburgh* (Edinburgh: Stewart, Ruthven and Co., 1792).
Watson, Hewett Cottrell, *Outlines of the Geographical Distribution of British Plants; Belonging to the Division of Vasculares or Cotyledones* (Edinburgh, 1832).
Watson, Hewett Cottrell, *Remarks on the Geographical Distribution of British Plants; Chiefly in Connection with Latitude, Elevation, and Climate* (London: Longman, Rees, Orme, Brown, Green, and Longman, 1835).
Watson, Hewitt Cottrell, *An Examination of Mr. Scott's Attack upon Mr. Combe's 'Constitution of Man'* (London: Longman, Rees, Orme, Brown, Green and Longman, 1836).

## SECONDARY SOURCES

Anderson, D., 'Scottish university professors, 1800–1939: Profile of an elite', *Scottish Economic and Social History* 7: 1 (1987), 27–54.
Anderson, R. D. *Universities and Elites* (Cambridge: Cambridge University Press, 1995).
Appel, Toby A., *The Cuvier–Geoffroy Debate: French Biology in the Decades before Darwin* (New York: Oxford University Press, 1987).
Baxter, Paul, 'Deism and development: Disruptive forces in Scottish natural theology', in Brown, Stewart J., and Fry, Michael (eds), *Scotland in the Age of the Disruption* (Edinburgh: Edinburgh University Press, 1993), 98–112.
Bourdier, Frank, 'Lamarck et Geoffroy Saint-Hilaire face au problème de l'évolution biologique', *Revue d'histoire des sciences*, 25: 4 (1972), 311–25.
Bowler, Peter J., *Darwin: The Man and his Influence* (Cambridge: Cambridge University Press, 1990).
Brock, W. H., 'Brewster as a scientific journalist', in Morrison-Low, A. D. and Christie, J. R. R. (eds), *'Martyr of Science' Sir David Brewster, 1781–1868* (Edinburgh: Royal Scottish Museum, 1984).
Brown, S. W., 'Smellie, William (1740–1795)', *Oxford Dictionary of National Biography* (Oxford: Oxford University Press, 2004; online edn, May 2008) <www.oxforddnb.com/view/article/25753> (last accessed 18 Oct 2012).
Browne, Janet, *Charles Darwin: Voyaging* (London: Pimlico, 1995).
Burkhardt, Richard W., Jr, 'The inspiration of Lamarck's belief in evolution', *Journal of the History of Biology* 5: 2 (1972), 413–38.
Burkhardt, Richard W., Jr, *The Spirit of System: Lamarck and Evolutionary Biology* (Cambridge, MA: Harvard University Press, 1977).
Burns, James, 'John Fleming and the geological deluge', *British Journal for the History of Science* 40: 2 (2007), 205–25.
Cantor, Geoffrey, *Quakers, Jews and Science: Religious Responses to Modernity and the Sciences in Britain, 1650–1900* (Oxford: Oxford University Press, 2005).
Centre National de la Recherche Scientifique, 'Liste des auditeurs du cours de Lamarck au Museum d'Histore Naturelle', <www.lamarck.cnrs.fr/auditeurs/liste.php?lang=fr&orig=Ecosse> (last accessed 27 October 2015).
Chitnis, Anand C., 'Medical education in Edinburgh, 1790–1826, and some Victorian social consequences', *Medical History* 17: 2 (1973), 173–85.
Cooter, Roger, *The Cultural Meaning of Popular Science: Phrenology and the Organization of Consent in Nineteenth-Century Britain* (Cambridge: Cambridge University Press, 1984).
Corsi, Pietro, 'The importance of French transformist ideas for the second volume of Lyell's *Principles of Geology*', *British Journal for the History of Science* 11: 3 (1978), 221–44.
Corsi, Pietro, *The Age of Lamarck: Evolutionary Theories in France, 1790–1830* (Berkeley, CA: University of California Press, 1988).

Corsi, Pietro, 'Jean-Baptiste Lamarck: From myth to history', in Gissis, Snait B. and Jablonka, Eva (eds), *Transformations of Lamarckism: From Subtle Fluids to Molecular Biology* (Cambridge, MA: MIT Press, 2011), 9–18.

Corsi, Pietro, 'The revolutions of evolution: Geoffroy and Lamarck, 1825–1840' (2011), <http://hsmt.history.ox.ac.uk/staff/documents/Corsi_Lamarckinthe1830s_Oct2011.pdf> (last accessed 6 May 2014).

Davie, George, *The Democratic Intellect* (Edinburgh: Edinburgh University Press, 2013).

Dean, Dennis R., 'Jameson, Robert (1774–1854)', *Oxford Dictionary of National Biography* (Oxford, Oxford University Press, 2004) <www.oxforddnb.com/view/article/14633> (last accessed 5 July 2012).

Desmond, Adrian, 'Robert E. Grant: The social predicament of a pre-Darwinian transmutationist', *Journal of the History of Biology* 17: 2 (1984), 189–223.

Desmond, Adrian, *The Politics of Evolution: Morphology, Medicine and Reform in Radical London* (Chicago, IL: University of Chicago Press, 1989).

Desmond, Adrian and Moore, James, *Darwin* (London: Penguin, 1991).

Desmond, Adrian and Parker, Sarah E., 'The bibliography of Robert Edmond Grant (1793–1874): illustrated with a previously unpublished photograph', *Archives of Natural History* 33: 1 (2006), 202–13.

Dobzhansky, Theodosius, 'Nothing in biology makes sense except in the light of evolution', *American Biology Teacher* (1973) 35: (3), 125–9.

Eddy, Matthew D., *The Language of Mineralogy: John Walker, Chemistry and the Edinburgh Medical School, 1750–1800* (Farnham: Ashgate, 2008).

Egerton, Frank N., *Hewett Cottrell Watson: Victorian Plant Ecologist and Evolutionist* (Aldershot: Ashgate, 2003).

Emerson, Roger, 'The contexts of the Scottish Enlightenment', in Broadie, Alexander (ed.), *The Cambridge Companion to the Scottish Enlightenment* (Cambridge: Cambridge University Press, 2003).

Eyles, V. A., 'Robert Jameson and the Royal Scottish Museum', *Discovery* 15: 4 (1954), 155–62.

Fyfe, Aileen, 'The reception of William Paley's *Natural Theology* in the University of Cambridge', *British Journal for the History of Science* 30 (1997), 321–35.

Fyfe, Aileen, *Steam-Powered Knowledge: William Chambers and the Business of Publishing, 1820–1860* (Chicago, IL: University of Chicago Press, 2012).

Glass, Bentley, Temkin, Owsei and Straus, William L., Jr, *Forerunners of Darwin* (Baltimore, MD: Johns Hopkins University Press, 1968).

Gray, James, *History of the Royal Medical Society, 1737–1937* (Edinburgh: Edinburgh University Press, 1952).

Gruber, Howard E., *Darwin on Man: A Psychological Study of Scientific Creativity* (London: Wildwood House, 1974).

Gruber, Jacob W., 'Who was the *Beagle*'s naturalist?', *British Journal for the History of Science* 4 (1969), 266–82.

Hodge, Jonathan, 'Darwin as a lifelong generation theorist', in Kohn, David (ed.), *The Darwinian Heritage* (Princeton, NJ: Princeton University Press, 1985).

Hodge, Jonathan, 'The notebook programme and projects of Darwin's London years', in Hodge, Jonathan and Radick, Gregory (eds), *The Cambridge Companion to Darwin*, 2nd edn (Cambridge: Cambridge University Press, 2009).

Hodge, Jonathan, 'On Darwin's science and its contexts', *Endeavour* 38: 3–4 (2014), 169–78.

Horn, David Bayne, *A Short History of the University of Edinburgh 1556–1889* (Edinburgh: Edinburgh University Press, 1967).

Jenkinson, Jacqueline, *Scottish Medical Societies, 1731–1939* (Edinburgh: Edinburgh University Press, 1993).

Jacyna, L. S., *Philosophic Whigs: Medicine, Science and Citizenship in Edinburgh, 1789–1848* (London: Routledge, 1994).

Jordanova, Ludmilla, *Lamarck* (Oxford: Oxford University Press, 1984).

Kaufman, M. H., 'John Barclay (1758–1826) extra-mural teacher of anatomy in Edinburgh: Honorary fellow of the Royal College of Surgeons of Edinburgh', *Surgeon* 4: 2 (2006), 93–100.

Laudan, Rachel, *From Mineralogy to Geology: The Foundations of a Science, 1650–1830* (Chicago, IL: University of Chicago Press, 1993).

Laurent, Goulven, 'Le cheminement d'Étienne Geoffroy Saint-Hilaire (1772–1844) vers un transformisme scientifique', *Revue d'histoire des sciences* 30: 1 (1977), 43–70.

Laurent, Goulven, 'Ami Boué (1794–1881): Sa vie et son œuvre', *Travaux du Comité Français d'Histoire de la Géologie*, troisième série T.VIII (1993) <www.annales.org/archives/cofrhigeo/ami-boue.html> (last accessed 27 April 2018).

Laurent, Goulven, 'Paléontologie(s) et évolution au début du XIXe siècle: Cuvier et Lamarck' *Asclepio* 52: 2 (2000), 133–212.

Lovejoy, Arthur O., *The Great Chain of Being* (Cambridge, MA: Harvard University Press, 1964).

Morrell, J. B., 'Practical chemistry at the University of Edinburgh, 1799–1843', *Ambix* 16 (1969), 66–80.

Morrell, J. B., 'The University of Edinburgh in the late eighteenth century: Its scientific eminence and academic structure', *Isis* 62: 2 (1971), 158–71.

Morrell, J. B., 'Science and Scottish university reform: Edinburgh in 1826', *British Journal for the History of Science* 6: 1 (1972), 39–56.

Morell, J. B., 'The Leslie affair: Careers, Kirk and politics in Edinburgh in 1805', *Scottish Historical Review* 54: 157 (1975), 63–82.

Morus, Iwan Rhys, 'Placing performance', *Isis* 101 (2010), 775–8.

Morus, Iwan Rhys, 'Worlds of wonder: Sensation and the Victorian scientific performance', *Isis* 101 (2010), 806–16.

O'Connor, Ralph, *The Earth on Show: Fossils and the Poetics of Popular Science, 1802–1856* (Chicago, IL: University of Chicago Press, 2007).

Ospovat, Dov, 'The Influence of Karl Ernst von Baer's embryology, 1828–1859: A reappraisal in light of Richard Owen's and William B. Carpenter's "Palaeontological application of "Von Baer's law"'', *Journal of the History of Biology* 9: 1 (1976), 1–28.

Ospovat, Dov, *The Development of Darwin's Theory: Natural History, Natural Theology and Natural Selection 1838–1859* (Cambridge: Cambridge University Press, 1981).

Pickstone, John V., *Ways of Knowing: A New History of Science, Technology and Medicine* (Manchester: Manchester University Press, 2000).

Rae, Isobel, *Knox: The Anatomist* (Edinburgh: Oliver & Boyd, 1964).

Rehbock, Philip F., *The Philosophical Naturalists: Themes in Early Nineteenth-Century British Biology* (Madison, WI: University of Wisconsin Press, 1983).

Richards, Evelleen, 'The "moral anatomy" of Robert Knox: The interplay between biological and social thought in Victorian scientific naturalism', *Journal of the History of Biology* 22: 3 (1989), 373–436.

Robertson, Forbes W., *Patrick Neil: Doyen of Scottish Horticulture* (Dunbeath: Whittles Publishing, 2011).

Rosner, Lisa, 'Barclay, John (1758–1826)', *Oxford Dictionary of National Biography* (Oxford: Oxford University Press, 2004) <www.oxforddnb.com/view/article/1345> (last accessed 5 July 2012).

Rostand, Jean, 'Étienne Geoffroy Saint-Hilaire et la tératogenèse expérimentale', *Revue d'histoire des sciences et de leurs applications* 17: 1 (1964), 41–50.

Rudwick, Martin, *Bursting the Limits of Time: The Reconstruction of Geohistory in the Age of Revolution* (Chicago, IL: University of Chicago Press, 2005).

Rudwick, Martin, *Worlds before Adam: The Reconstruction of Geohistory in the Age of Reform* (Chicago: University of Chicago Press, 2008).

Schaffer, Simon, 'The nebular hypothesis and the science of progress', in Moore, James R. (ed.), *History, Humanity and Evolution: Essays for John C. Greene* (Cambridge: Cambridge University Press, 1989).

Scull, Andrew, 'Browne, William Alexander Francis (1805–1885)', *Oxford Dictionary of National Biography* (Oxford University Press, 2004; online edn, September 2010) <www.oxforddnb.com/view/article/46958> (last accessed 5 July 2012).

Secord, James A., 'Beyond the veil: Robert Chambers and *Vestiges*', in Moore, James R. (ed.), *History, Humanity and Evolution: Essays for John C. Greene* (Cambridge: Cambridge University Press, 1989), 165–94.

Secord, James A., 'Edinburgh Lamarckians: Robert Jameson and Robert E. Grant', *Journal of the History of Biology* 24: 1 (1991), 1–18.

Secord, James A., *Victorian Sensation: The Extraordinary Publication, Reception, and Secret Authorship of Vestiges of the Natural History of Creation* (Chicago, IL: University of Chicago Press, 2003).

Secord, James A., *Visions of Science: Books and Readers at the Dawn of the Victorian Age* (Oxford: Oxford University Press, 2014).

Shapin, Steven, 'Phrenological knowledge and the social structure of early nineteenth-century Edinburgh', *Annals of Science* 3 (1975), 219–43.

Shapin, Steven, *A Social History of Truth: Civility and Science in Seventeenth-Century England* (Chicago, IL: University of Chicago Press, 1994).

Shapin, Steven, *The Scientific Life: A Moral History of a Late Modern Vocation* (Chicago, IL: University of Chicago Press, 2007).

Sher, Richard B., *Church and University in the Scottish Enlightenment: The Moderate Literati of Edinburgh* (Edinburgh: Edinburgh University Press, 2015).

Sloan, Phillip R., 'Darwin's invertebrate program, 1826–1836: Preconditions for transformism', in Kohn, David (ed.), *The Darwinian Heritage* (Princeton, NJ: Princeton University Press, 1985).

Sloan, Phillip R., 'Darwin, vital matter, and the transformism of species', *Journal of the History of Biology* 19: 3 (1986), 369–445.

Sloan, Phillip R., 'The making of a philosophical naturalist', in Hodge, Jonathan and Radick, Gregory (eds), *The Cambridge Companion to Darwin*, 2nd edn (Cambridge: Cambridge University Press, 2009).

Sulloway, Frank J., 'Darwin and his finches: the evolution of a legend', *Journal of the History of Biology*, 15: 1 (1982), 1–53.

Sweet, Jessie M., 'Robert Jameson's Irish journal, 1797', *Annals of Science* 23: 2 (1967), 97–126.

Sweet, Jessie M. and Waterston, Charles D., 'Robert Jameson's approach to the Wernerian theory of the earth, 1796', *Annals of Science* 23: 2 (1967), 81–95.

Taylor, Clare L., 'Knox, Robert (1791–1862)', *Oxford Dictionary of National Biography* (Oxford University Press, 2004) <www.oxforddnb.com/view/article/15787> (last accessed 5 July 2012).

Topham, Jonathan R., 'Science and popular education in the 1830s: The role of the Bridgewater Treatises', *British Journal for the History of Science* 25: 4 (1992), 397–430.

Topham, Jonathan R., 'Beyond the "common context": The production and reading of the Bridgewater Treatises', *Isis* 89: 2 (1998), 233–62.

Topham, Jonathan R., 'Science, print and crossing borders: Importing French science books into Britain, 1789–1815', in Livingstone, David N. and Withers, Charles W. J. (eds), *Geographies of Nineteenth-Century Science* (Chicago, IL: University of Chicago Press, 2011).

Topham, Jonathan R., 'The scientific, the literary, and the popular: Commerce and the reimagining of the scientific journal in Britain, 1813–1825', *Notes and Records of the Royal Society* 70: 3 (2016), 305–24.

Van Wyhe, John, '"My appointment received the sanction of the Admiralty": Why Charles Darwin really was the naturalist on HMS Beagle', *Studies in History and Philosophy of Biological and Biomedical Sciences* 44 (2013), 316–26.

Warren, John S., 'Darwin's missing links', *History of European Ideas* 43: 8 (2017), 929–1001.

Withers, Charles W. J., 'Towards a history of geography in the public sphere', *History of Science* 37: 1 (1999), 45–78.

Yeo, Richard, 'Science and intellectual authority in mid-nineteenth-century Britain: Robert Chambers and *Vestiges of the Natural History of Creation*, *Victorian Studies* 28: 1 (1984), 5–31.

# Index

Ainsworth, William Francis, 63, 66, 68, 69, 186
atmosphere, composition of the, 100, 101, 103, 117–18, 136, 137, 144, 149

Baconian (inductive) method, 114–15, 119, 183–4, 191, 200–1
Baer, Karl Ernst von, 164–5, 170, 176
Barclay, John, 25, 29–30, 32, 54–5, 58, 65, 70, 133, 134, 136
Barry, Martin, 168, 170
Blainville, Henri Marie Ducrotay de, 30, 58, 138, 139, 140
Blumenbach, Johann Friedrich, 148
Bonnet, Charles, 39, 42, 44
Boué, Ami, 67, 84, 85, 88–9, 91, 94, 101
Brewster, David, 16, 18, 46, 52, 58, 66, 67, 76, 130, 132, 158, 159, 160, 183, 199, 200
  Editor of *Edinburgh Journal of Science*, 66, 67, 126, 186, 199
Brougham, Henry, 20, 23
Browne, William A. F., 61–3
Buckland, William, 101, 132, 157, 163
Buffon, George Louis Leclerc, Comte de, 38, 40, 41–2, 44, 78, 87, 114, 118, 119, 130, 140, 143, 144, 148, 180, 197
  *Histoire naturelle*, 38, 40, 41, 87
  internal moulds, theory of, 87, 118
  organic molecules, theory of, 44, 87, 130, 140, 180

Cambridge, University of, 2, 11, 12, 13, 15, 20–1, 22, 32, 186
Candolle, Alphonse de, 85
Carpenter, William Benjamin, 24, 168, 170–2, 176–7, 185, 200
catastrophism, 99, 101, 114, 128, 157–8, 161, 190, 191
Chalmers, Thomas, 130

Chambers, Robert, 161–2
  *Vestiges of the Natural History of Creation*, 67, 75–6, 156, 158, 159, 160, 161–77, 183, 198, 199–200
Cheek, Henry H., 52–4, 61, 64, 66, 70, 141–8, 161, 186, 189, 190, 191, 199
  Editor of *Edinburgh Journal of Natural and Geographical Science*, 52, 66, 68–9, 70, 142, 143, 148, 186
Church of England, 10, 12, 13, 22
  Thirty-Nine Articles, 2, 20, 21
Church of Scotland, 8, 9, 21, 38, 52, 161
  Disruption, the, 132, 159, 160, 161
  Evangelical Party, 18, 130–1, 158–9
  Moderate Party, 9–10
climate, influence of, 41–2, 56, 77–8, 82–3, 84, 89, 90, 91, 92, 93, 98, 100, 110, 120, 122, 144, 145, 180, 181
Cockburn, Henry, 15, 18, 19
Coldstream, John, 59, 61, 63
Combe, George, 55, 62, 164, 172–7, 179
  *Constitution of Man*, 164, 172, 173–4, 176
Conybeare, William, 157
Cullen, William, 14, 38
Cunningham, Robert James Hay, 85–6
Cuvier, Georges, 3, 48, 55, 58, 66, 119, 125, 141–2
  *Essay on the Theory of the Earth*, 48, 89, 121, 158

Darwin, Charles, 1, 5, 25, 48, 59, 92, 98, 182–90, 200–1
  Essay of 1844, 184, 185, 188
  and finches, 184
  notebooks, 157, 184, 185, 187, 189, 191, 201
  *The Origin of Species*, 156, 184–5, 189, 190

Darwin, Charles (*cont.*)
  voyage of the *Beagle*, 157, 184, 185, 186, 187, 189, 191, 200
Darwin, Erasmus, 65, 102, 182, 189
Deluge, the, 38, 89
divine intervention, 130, 131, 135, 158
dogs, breeds of, 41–2, 90
Duncan, James, 159–60

Edinburgh, University of, 1–6, 7–33, 196–7
  conflict with Edinburgh City Council, 11–12
  continental connections, 3, 200
  English students at, 21–2, 23, 27
  medical school, 23–8
  natural history syllabus, 8, 49
  nepotism, 17, 26
  patronage, 18–19
  remuneration of professors, 15
  secularism, 21
  students, 12–14, 19
  teaching, 14–16
Egypt, Napoleon's expedition to, 116, 121, 125, 139
embryology, 115, 116, 164–5, 168–9, 172, 176–7, 178–9
Esmark, Jens, 67, 84–5
extinction, 76, 89, 99, 103, 114, 124–5, 128, 144
extra-mural anatomy schools, 2, 28–32, 54–9, 126, 128–9, 135–6, 140–1, 198

Firth of Forth, 2, 46, 59, 92, 126
Fleming, John, 3, 58, 60, 66, 68, 75–6, 83–4, 102, 130, 131, 132–5, 160–1
  *Philosophy of Zoology*, 83, 86–7, 134, 135, 160, 166
Fletcher, John, 58–9, 94–5, 101–2, 140–1
  *Rudiments of Physiology* 58–9, 129–30, 166, 168, 169, 170, 176
Forbes, Edward, 48
fossil record, 82–9, 91–2, 94, 98, 99–103, 114, 116–17, 122, 125, 128, 138, 139, 157, 179, 197
Free Church of Scotland, 75, 132, 159, 160, 161
French Revolution, 19, 198, 200

Geoffroy Saint-Hilaire, Étienne, 3, 30, 55, 58, 100–1, 103, 115–20, 141–2, 188
  *Anatomie Philosophique*, 115

  and crocodiles, 101, 116, 139
  experimental transformism, 117, 143, 187
  and ibises, 125
Glasgow, University of, 11, 14, 120, 162
Grant, Robert, 2, 33, 56–9, 66, 67, 70, 88, 91, 92–4, 95, 98–9, 101, 102, 126–8, 135, 136–8, 145, 156, 161, 185–7, 189, 198–9
great chain of being, 39–40, 42, 99–100
Grierson, James, 131, 159

Hall, James, 60, 79–80
Herschel, John, 183
Home, Henry, Lord Kames, 2, 7, 9, 38
Hope, Thomas Charles, 11, 15, 25, 60, 80
Hume, David, 2, 7, 9, 18
Hutton, James, 46, 65, 78, 79, 96
hybridisation, origin of new species by, 42–3, 167, 179, 180

inheritance of acquired characteristics, 4, 110–11, 129, 147–8

Jameson, Robert, 1–2, 25, 30, 33, 37, 44–54, 59–69, 75, 80, 82–3, 89–91, 99–100, 102, 121–3, 135–6, 161
  Editor of Cuvier's *Theory of the Earth*, 48, 89, 121, 157–8
  Editor of *Edinburgh (New) Philosophical Journal*, 30, 45, 46, 52, 66–7, 70, 84. 85, 86, 88, 91, 135–6
  Keeper of the Edinburgh College Museum, 14, 30, 44, 46, 48, 49–52, 53–4
  patronage network, 45–6, 58
  teaching, 49–51
Jardin des Plantes, 107, 189
Jardine, George, 14
Jardine, William, 68, 159

Knox, Robert, 2, 29–32, 33, 55–6, 66, 67, 70, 123, 128–9, 138–40, 156, 177–9

Lamarck, Jean-Baptiste, 4, 65, 90, 95–101, 108–15, 125
  *Histoire naturelle des animaux sans vertèbres*, 102, 112, 120, 123, 126, 132, 134, 148, 159

*Hydrogéologie*, 96–7, 103, 137, 149, 197
*Philosophie zoologique*, 110, 112–13, 120, 124, 134, 147, 148, 166
  Scottish students of, 120
Leslie, John, 7, 11, 16
  'Leslie affair', 17–18
Linnaeus, Carl, 42–3, 90, 91, 102, 179, 180, 197
Lord, Percival Barton, 168, 169
Lyell, Charles, 81, 159, 189

MacGillivray, William, 46, 66, 68, 123
Mackenzie, George, 80
Maillet, Benoît de, 65, 77
Malthus, Thomas, 157, 187, 188, 191
materialism, 3, 4, 55, 59, 61, 62–3, 108, 128, 132–3, 134, 196, 199
Meckel, Johann Friedrich, 107, 148, 169, 172
Miller, Hugh, 131, 158, 159, 160, 200
moderate literati, 9–10, 32
Monro, Alexander, primus, 16, 25
Monro, Alexander, secundus, 25, 41, 42
Monro, Alexander, tertius, 24, 25, 26, 28, 29, 58
monsters, 117, 143, 165
Muirhead, Lockhart, 120–1, 124, 125, 131, 135
Muséum d'Histoire Naturelle, 103, 107, 111, 114–15

natural theology, 8, 18, 176, 196, 200
Neill, Patrick, 53, 65, 68, 69, 70
Newtonianism, 79, 108
Nichol, John Pringle, 162, 163
*Nisus formativus*, 118

Oken, Lorenz, 107
Owen, Richard, 8, 24, 185
Oxford, University of, 2, 3, 12, 13, 15, 20–1, 22, 32

Paley, William, 196
phrenology, 55, 61, 62, 63, 172–5
Playfair, John, 65, 80, 82
Plinian Natural History Society, 4, 58, 59–64, 70, 102, 123, 141, 147, 175, 179, 185–6, 188, 190

race, theories of, 30, 56, 64–5, 129, 143
Ramsay, Robert, 38

Robertson, William, 3, 7, 9–10, 19, 21, 196
Royal Medical Society, 20, 29, 47, 64, 81, 83, 143, 175, 186
Royal Physical Society, 20, 64–5, 81
Royal Society of Edinburgh, 4, 38, 58, 80, 132
Royal Society of London, 4

Scott, Walter, 4, 13, 60
Scottish Universities Commission (1826), 11, 13, 14, 15–16, 17, 19, 21–2, 23, 26, 27–9, 49, 51–3, 54, 60, 80
Sedgwick, Adam, 101, 157, 163
Serres, Étienne, 115, 116, 137, 164, 169, 172, 178
Smellie, William, 37, 39–40, 41
Smollet, Tobias, 8–9
species concept, 111, 134, 143–4, 178, 181
sponges, 3, 58, 67, 92–3, 126, 127, 136, 199
spontaneous generation, 43–4, 61, 109, 112–13, 119, 121, 124, 129, 163, 167
Spurzheim, Johann, Gaspar, 55, 173
Stewart, Dugald, 7, 10
subtle fluids, 97, 108–10, 111

Tiedemann, Friedrich, 107, 127, 130, 164
transcendental anatomy, 56, 58, 70, 107, 115, 119, 135, 137–9, 140, 141–3, 149, 164, 167, 168–72, 176, 188, 190–1, 198

Uniformitarianism, 79, 81, 96–7, 137, 149, 197
University College London, 11, 12, 58, 137–8

variation, causes of, 118, 144–7, 187–9
varieties, 41–3, 90, 120–2, 143–4, 179, 180–1, 188
Virey, Julien Joseph, 129, 133, 165
vital principle, 54–5, 61, 63, 87, 108, 133, 134, 199

Walker, John, 6, 10, 21, 37–44, 46, 47, 48, 49, 91, 102, 121, 122
Watson, Hugh Cottrell, 8, 156, 179–82
Werner, Abraham Gottlob, 3, 38, 46–7, 65, 75, 76, 77, 78, 79, 83, 90, 107, 197

Wernerian Natural History Society, 30, 45, 52–3, 54, 65–6, 68, 69, 131, 136, 149, 159

*Memoirs of the Wernerian Natural History Society*, 46, 65, 85, 127, 131, 159

EU representative:
Easy Access System Europe
Mustamäe tee 50, 10621 Tallinn, Estonia
Gpsr.requests@easproject.com

www.ingramcontent.com/pod-product-compliance
Lightning Source LLC
Chambersburg PA
CBHW070349240426
43671CB00013BA/2453